WE SEVEN

BY THE ASTRONAUTS THEMSELVES.

M. Scott Carpenter   L. Gordon Cooper, Jr.

John H. Glenn, Jr.   Virgil I. Grissom

Walter M. Schirra, Jr.   Alan B. Shepard, Jr.

Donald K. Slayton

Simon & Schuster Paperbacks
New York   London   Sydney   Toronto

Simon & Schuster Paperbacks
A Division of Simon & Schuster, Inc.
1230 Avenue of the Americas
New York, NY 10020

This Simon & Schuster trade paperback edition January 2010

SIMON & SCHUSTER PAPERBACKS and colophon are
registered trademarks of Simon & Schuster, Inc.

For information about special discounts for bulk purchases,
please contact Simon & Schuster Special Sales at
1-866-506-1949 or business@simonandschuster.com.

The Simon & Schuster Speakers Bureau can bring authors
to your live event. For more information or to book an event,
contact the Simon & Schuster Speakers Bureau at
1-866-248-3049 or visit our website at www.simonspeakers.com.

Manufactured in the United States of America

10   9   8   7   6   5   4   3   2   1

Library of Congress Catalog Card Number: 62-19074

ISBN 978-1-4391-8103-4

CONTENTS

At a press conference in Washington in 1959, at which they were first introduced to the public, the seven Astronauts all raise their hands when a reporter asks who is ready to go into space then and there. They are, from the left: Slayton, Shepard, Schirra, Grissom, Glenn, Cooper and Carpenter. Schirra and Glenn doubled their vote by raising both hands.

THOSE SEVEN—
AN INTRODUCTION

This book, like the man-in-space program which it describes, is the result of a considerable amount of teamwork. Its authors, the seven Astronauts, took turns carrying out many of the duties involved in Project Mercury; and they have followed the same procedure in the preparation of this volume. As the reader will soon discover, it is a personal narrative, full of suspense and adventure, reminiscences and beliefs, facts and opinions, good days and bad days, and patiently detailed descriptions of a number of complex technical matters which the Astronauts themselves—as the men who know the subject best—are best equipped to explain to others.

It has been the privilege of Life's *staff, including the editor of this book, to work closely with the Astronauts since they first joined Project Mercury in the spring of 1959. After this introduction, the editor will speak only occasionally, with brief notes and in this same typeface, to clarify points which might not be evident in the first-person accounts. But before John Glenn takes up the narrative, the reader may be interested in some additional background about the authors which they have not included in their own chapters, either out of modesty or because none of them would be so presumptuous.*

Indeed, one of the most fascinating aspects of Project Mercury has been the sociology of the Astronauts themselves. What kind of man could manage to be part pilot, part engineer, part explorer, part scientist, part guinea pig—and part hero—and do equal justice to each of the diverse and demanding roles that was being thrust upon him? What kind of man would be strong enough physically to sustain the tremendous stresses and strains of a flight through space, and then be wise enough—and strong enough in other ways—to sustain the pressures of public adulation when he returned home? And what kind of man, above all, would be best qualified to help set the rare standards of courage and stamina, skill and alertness, vision and intelligence that would be needed to lead him and his colleagues to the moon and to Mars and to distant places beyond?

The National Aeronautics and Space Administration (NASA) was pondering these same questions in the fall of 1958 when it first began to lay the groundwork for its man-in-space program and had to decide, among many other pressing problems, exactly what kind of human being it needed to man the cockpit. NASA arrived first at some general conclusions:

The men would have to be daring and courageous—and offer some proof of this in their records. It was obvious that space, with all of its unknowns—from plunging meteorites to searing radiation—was a hostile environment, and that a journey through it in a new kind of craft that no one had ever flown was bound to be hazardous.

Courage in itself, however, was not enough. The men would also have to be the kind who would remain cool and resourceful under pressure. For if they were to guide a new machine through a hostile environment, they would be faced with emergencies that no one could foresee; and no matter how cleverly the system was constructed to get them there and back, there would be moments when it could fail and the men would be the masters of their own destiny.

The men would also have to be physically strong, of course. For space flight would be strenuous work; it would expose them to greater stresses than most pilots had ever encountered, even in combat. And the men would have to have nerves of steel. They would have to be devoid of emotional flaws which could rattle them or destroy their efficiency when they found themselves in a crisis.

Taking all of these problems into account, and adding whatever bits and pieces of knowledge it could extrapolate from the detailed plans it was already drawing up for some of the equipment the men would use, NASA put together a list of specific requirements to describe the kind of men it was looking for:

They would have to be young enough to be in their physical prime— with a maximum of body reserves to draw on—and yet mature enough to have lost the rash impulses of youth. The top age was set at forty. NASA felt that a man who was older than that might be otherwise qualified for the rigors of a flight or two into space, but that by the time he was fully trained he would be approaching the point where his long-range useful- ness as a pilot would be limited. (Since then, the age limit for new Astro- nauts entering the program has been lowered to thirty-five.)

Their maximum height was set at 5 feet, 11 inches. The reason for this arbitrary cutoff, which weeded out a number of disappointed and other- wise well-qualified six-footers, was equally practical. The capsule was already on the drawing boards, and its dimensions were dictated by the diameter of the available boosters—the Redstone and Atlas missiles—which would launch it into space. It was, to be precise, 74 inches wide at the base. This meant that once he had donned his bulky pressure suit, bulging helmet and pressurized boots, a man who stood any taller than 71 inches to begin with would simply not be able to squeeze in for the ride.

The weight limit of each man was set at 180 pounds. There were two reasons for this stipulation. One was that weight was a major factor in the over-all payload. The boosters were limited in their power, and each pound of payload that the man himself displaced left that much less room for the equipment which he would need to keep him alive and bring him home. A more important reason, however, was that an Astronaut who met the height requirement but weighed more than 180 pounds was probably overweight and would therefore bring along with him a metabolic and circulatory system that would put him at a distinct disadvantage as he tried to cope with such physical stresses of space flight as prolonged weightlessness and rapid and severe changes in temperature.

NASA also announced that no applicant would be considered for the team who had not earned either a formal engineering degree or its equivalent. The reason for this requirement lay in the basic philosophy of the project itself. Since it was starting from scratch, Mercury needed all of the hardheaded engineering experience it could muster from the very beginning, and the Astronauts who were destined to fly the new equipment were expected to pitch in from the start and articulate their own suggestions, in precise, nuts-and-bolts terms, with the project engineers who were designing and building it.

In addition to all this, NASA narrowed its search almost immediately to the ranks of practicing military test pilots. The theory here was that the only men who already possessed the kind of training required to master such a complex and fast-flying machine as the space capsule—and had developed an unfailing instinct for making calm, steely, split-second decisions at high speeds and high altitudes—were the same men who had helped to test out the nation's fastest and hottest military aircraft. It was a case of picking the roundest pegs available to fit round holes.

Finally, with all of these criteria in mind, NASA went hunting for volunteers who would fill the bill. "What we're looking for," said an Air Force general—who made it clear he was only half kidding—"is a group of ordinary supermen."

The first flopping of the personnel cards turned up the names of 508 young test pilots who met the basic requirements—at least on paper. Then, with the help of suggestions from the commanding officers who had employed these pilots and were familiar with their everyday work, the list was narrowed to 110.

This group was later pared down to 69, partly on the basis of evaluations of the men which were supplied by instructors who had taught the pilots how to fly and knew at first hand the quality of their nerves and their reflexes. The men who still remained on the list after this culling out were summoned to Washington for a series of interviews with NASA representatives who filled them in on what Project Mercury was all about and asked them if they were at all interested in trying out for the team. At this point, a total of 37 men—most of whom were simply not willing or eager to make such a radical change in their careers—dropped out of the running. The 32 who volunteered to take a chance went on to receive an exhaustive series of medical examinations and psychological tests— which the Astronauts will explain in greater detail in the first two chapters. Another 14 men fell by the wayside as a result of this sifting-out procedure. This left 18 still under consideration. NASA made its final choice from this group, and, at 2 P.M. on April 9, 1959, the seven men who made the grade were paraded onto a stage in Washington, D. C., and introduced to the world at a bustling press conference. The reporters who were present thought up a few questions, and when one of them asked for a show of hands of any of the Astronauts who felt they were ready to go off into space then and there, all seven raised their hands. The adventure was on.

At first glance, there were some striking similarities among the seven. All of them were married and had children. All hailed from small towns and cities. All had brown hair except one—John Glenn, whose thinning hair was red. There was a difference of only four inches in their height (Shepard, the tallest, was right on the nose at 5 feet, 11 inches; Grissom, the shortest, was 5 feet, 7 inches; Carpenter, Glenn and Slayton stood at 5 feet, 10½ inches; Schirra was 5 feet, 10 inches; and Cooper was 5 feet, 9 inches). Three of them—Cooper, Shepard and Slayton—had blue eyes. Grissom's and Schirra's were brown. Glenn's and Carpenter's eyes were green. Their ages when they joined Mercury ranged from 32 (Cooper) through 33 (Grissom), 34 (Carpenter), 35 (Shepard and Slayton) and 36 (Schirra) to 37 for Glenn. The youngest man, Cooper, was also the lightest—150 pounds. Grissom, the shortest man, was a compact 155. Shepard, Slayton and Carpenter each weighed 160. John Glenn's weight stood right at the maximum—180. And Wally Schirra had to slim down from 185 to make the team. (Both Glenn and Schirra have had to watch their weight closely ever since in order to keep in shape. On the morning of his orbital flight, Glenn weighed 171 pounds, 7 ounces.)

The major common denominator, of course, was that all seven men were experienced flyers. Three of them had started their military flying in time to be combat veterans. Glenn had won the Distinguished Flying Cross five times and an Air Medal with eighteen clusters for service in World War II and Korea. Grissom had been awarded the Distinguished Flying Cross and the Air Medal with cluster for combat missions in Korea. Schirra had won the Distinguished Flying Cross and Air Medal with cluster, also for service in Korea.

This was the pattern—seven men with similar background and technical education, the same kind of skills and know-how, and only small variations in size, shape and coloring to distinguish them one from the other. Since they had all been pushed through the same fine sieve of interviews and tests, one might even have expected that they would come tumbling out of the hopper like seven peas in a pod. But the Astronauts did not turn out like that at all. On the contrary, NASA wound up with a team of seven distinctly original personalities and a rich mixture of private attitudes, personal characteristics and professional ideas that were to prove invaluable in the day-to-day workings of the program.

"There's no doubt about it," Alan Shepard told a friend of his one day, "we are seven different individuals, seven different personalities. We all have different strengths and abilities and different temperaments. But I think we balance each other out pretty well. Some of us are stronger in certain fields. Some have stronger personalities. Some have a moderating influence. Nobody pulls any punches when we get together. But I guess one reason the psychiatrists went over us so thoroughly was to find people who would work together. This is a really fine opportunity to show that unification can work—on our level, anyway."

It was typical of Alan Shepard that he would make that particular point. For Shepard, the bright-eyed and articulate U.S. Navy Commander from New Hampshire, has won a reputation for being an expert mediator on the many committees which have carried out much of Mercury's work behind the scenes. The Astronauts, who are an extremely competitive lot and have learned to rate each other's contribution to the team realistically, rank Shepard high on their own private lists. "He is very bright," says one of a very bright group. "He is an excellent pilot," says one of the best pilots in the business. "The Navy lost a good officer when Al came into the program," another Astronaut has said. "He takes the admiral's big-picture view of the fleet. He feeds all the little items into place and is not confused by the details. Once he sets himself to an argument, he's tough and firm—and usually right."

Shepard has a sharp wit, an ingratiating but well-modulated charm and an excellent sense of humor. It was Al who introduced José Jiménez into the Astronaut clan. José, the brainchild and alter ego of comedian Bill Dana, is a quixotic Spanish Astronaut with a hilarious accent and a wispy, "please-don't-send-me" attitude who apparently struck Shepard's funny-bone just right. Al found the Jiménez brand of humor such a handy device for relaxing the troops when tension was building up before a flight that he once arranged for a tape of some of José's dialogue to be played in the Mercury Control Center during a practice run. And—as later passages in the book will show—he has repeated the Jiménez theme on appropriate occasions. But, when the chips are down, Shepard is all business. At the same time he was helping to prepare for Glenn's orbital mission, Shepard was also given another vital assignment: to test out designs for future spacecraft and some of NASA's long-range plans for missions to the moon. His favorite diversions are those of a natural and graceful athlete—golf, water-skiing and ice skating. He is also so fond of a white Corvette sports car that he has been known to drive it the 800 miles from his home to Cape Canaveral rather than fly.

Gus Grissom, the compact young Air Force captain from Michigan who followed Shepard into space, may be the shortest of the Astronauts, but he is long on engineering savvy and flying experience. And though Gus had to bear the embarrassment of losing his own capsule in a freak accident soon after it hit the water, the flight he made was so successful that no further Redstone missions were called for, and John Glenn's orbital mission was brought that much closer. (Gus's colleagues, incidentally, are convinced that the loss of his capsule was not—and could not have been—Gus's fault. The latest evidence, if any is needed, were the two bruised knuckles which John Glenn received when he blew the hatch on his own capsule. The plunger snapped back at Glenn just as it

would have at Grissom had he actually hit it, even inadvertently. Grissom
suffered no such wounds.)

Gus is a little bear of a man. He is fond of hunting and fishing, and he
is also an excellent handball player. In one series of games between the
Astronauts, the only man who managed to beat Gus was Alan Shepard.
There were rumors at the time that Gus blew the game because he knew
that Al was anxious to win. This was unlikely, however. The Astronauts
compete in nearly everything they do. Competition comes naturally to
them, and it would not be like Gus Grissom at all to let another Astronaut beat him at anything. He is a no-nonsense pilot, a steady, dedicated
professional who is completely absorbed in his work.

Underneath this working façade, Gus Grissom is a warm, friendly and
extremely thoughtful man. He is a country boy at heart, and he would
obviously much rather be working or spending his spare time with his
two young sons than riding in parades or going to parties. He scoffs at the
idea that just because he is an Astronaut he might be expected to play a
role or dart around on public view in the goldfish bowl. He has also re-
frained from acting like an Astronaut at home. "Betty is interested in
me," he has said. "If I have any homework to do, I try to wait until the
boys are in bed so they won't think I'm showing off." Gus was completely indifferent to the spotlight which each of the Astronauts found
pointed at him as he completed a flight. "Most people who know me," he
has said, "know I'm not the hero type. Personal prestige I couldn't care
less about." After his own flight, as a matter of fact, Grissom went to
some length to shun the limelight rather than bask in it. He wanted his
family to get back to a normal life and he was anxious to get back to
work. The reporters and the photographers, however, were still dogging
him. One day, when Gus was scheduled to board a commercial plane in
Orlando, Florida, to visit an Air Force installation in Texas, he put his sly
sense of humor to work and tried out a disguise consisting of a straw hat
and a pair of dark glasses. "How do I look?" he asked Deke Slayton,
hopefully. "You look just like Gus Grissom in dark glasses and a hat,"
Slayton told him. Undaunted, Grissom proceeded to Orlando with disguise in place and slipped right past a group of reporters and photographers who had staked out the airport. He considered this little victory a
major triumph of privacy over publicity.

John Glenn, the senior man on the team in terms of age and previous fame—he set a transcontinental speed record in 1957—is the personification of still another style. Sternly self-disciplined and almost ascetic in his pursuit of perfection, John has demonstrated an air of complete and absolute dedication to his job. He has also portrayed the most consciously thought-out image of what an "Astronaut" should be and how he should behave—both in public and in private—of any of his colleagues. It is John's firm conviction that the Astronauts are not simply seven experienced test pilots who have been chosen to extend the frontiers of this particular profession, but that they are the first of a new and even heroic breed of men who have the enormous responsibility of serving as symbols of the nation's future. "John tries to behave," a friend of his has said, "as if every impressionable youngster in the country were watching him every moment of the day."

John has spared no one, least of all himself, in this idealistic pursuit. Though his personal life centers completely around his attractive family, Glenn was the only one of the seven who did not arrange for his family to live within commuting distance of Project Mercury headquarters at Langley Air Force Base in Hampton, Virginia. Instead, he chose to keep his wife, Annie, and two teen-age children in the comfortable house he owned outside Washington and to take a room for himself in the bachelor officers' quarters at Langley. One reason for this was that John wanted to keep his children settled in the good schools which they were already attending. But there was also another reason: he felt it would be best for his training schedule if he had no distractions. Annie agreed with him. "I think it was good for John to be alone," she has said, "to work out his studies and get a good night's rest. This way he doesn't have the worries of when to order more wood or fix the front door." Even when he did get home on weekends, John tended to live his work. He felt that his family should know everything they could about his job so they would worry less about the dangers involved in it. Family conversations often centered about the training experiences he had had during the week.

Despite this relentless sense of dedication, John Glenn is a warm, convivial and sincerely friendly human being. "He is a very sentimental man," says Annie. "Good music makes him cloud up." He is also a good talker, an amusing storyteller, and a jovial party tenor. But through it all there runs a streak of moral conviction and public responsibility that will not lie. On the first night that John Glenn got home to Arlington after his orbital flight, he was dead tired. He had just come from a long day of parades, handshaking, noisy police escorts, a reception at the White House and a moving speech before a crowded session of Congress. Through it all John Glenn had portrayed the perfect image of the modest, dedicated and patriotic hero. He had probably done more in this one day than dozens of other people could have done in months to sell the U.S. space effort to Congress and to the nation. And now the day was over. But there was no chance to rest or to wind down. The house was crammed with bags of mail that would somehow have to be answered, stacks of gifts that would have to be acknowledged, a platoon of friends and relatives to say hello to, and a squad of NASA officials who had a number of problems to discuss—including plans and logistics for tumultuous trips later in that same week to New York and to New Concord, Ohio, his home town. And, though Annie had long since arranged for an unlisted number, the telephone was ringing constantly in the master bedroom. Finally, after pausing briefly to watch a portion of his Congressional speech on a TV news show, John tried to settle down for the long evening of work that lay ahead. But now there was another problem. Among the many presents that had been delivered in the family's absence was a huge crate of fresh eggs. There was no place in the house to store so many eggs, and John was sure they would spoil before he and his family could either eat them or find time to give them away. He could not bear to think of so much food going to waste, however, when there were undoubtedly people in the city who could use it. And so, everything came to a stop in the Glenn household that evening until someone thought of a solution. A friend who dropped by suggested that perhaps Glenn's church would know what to do with the eggs and that someone might call the minister to see if he could help. This was done, and then John Glenn was able to relax a little and go back to solving some of the other problems of being an Astronaut and a hero.

There are more profound differences between any two of the Astronauts than are apparent in a simple comparison of their individual per-

sonalities or physical appearance. Scott Carpenter and Deke Slayton, for example (who happen to be of the same height and weighed the same when they joined the program), have found themselves on different sides of the fence on a number of occasions when a matter of training or project policy was at stake. These differences of opinion have been healthy and positive, and more often than not they have resulted in constructive ideas for the program. One thing Deke and Scott did not agree on was the value of the three-axis control system which had been designed and slated for the capsule before they joined the project and which the Astronauts will describe in detail later in the book. The point of the disagreement was that pilots had traditionally used a two-axis control stick in aircraft—to control the plane's attitude in pitch and roll—and had used their feet to manage the third, or yaw, axis. The Mercury engineers decided at the start of the program that all three control functions should be centralized in one stick which the Astronaut would manipulate with his right hand. His feet would have no control function at all. This was fine with Carpenter. Scott felt that the Astronauts' mission was a decided departure from traditional flying and that it was entirely fitting for the new art of space flight to make use of a brand-new system of controls. It was not fine, however, with Slayton. A pilot who would rather fly an airplane than eat—unless, perhaps, the menu includes fresh oysters, of which he has been known to down two or three dozen at one sitting—Deke was partial to the old system of controls. He felt that the change was unnecessary, and that it would also involve extra training for the Astronauts and a consequent loss of time before they could become accustomed to it. Neither Deke nor Scott looked at the change from a purely mechanical point of view. Each man's personal philosophy of flight was involved in his attitude towards it. And neither man was alone in his argument. The other Astronauts also took sides on the matter—with the exception of Shepard, who, typically enough, saw the merits on both sides and took a position more or less in the middle.

Like all of the others, Deke Slayton has a style all his own. A taciturn and somewhat shy man with piercing blue eyes that crinkle around the edges and merge into a lined and leathery "pilot's face," Deke looks on first sight as if he might have stepped out of a Steve Canyon *or* Terry and the Pirates *comic strip—where he would play the part of the* quiet, solid executive officer. A farm boy from Sparta, Wisconsin, Slayton has a deep, vibrant voice with a touch of country drawl still embedded in it. He can use this voice, as his colleagues well know, to be a frank and stubborn man when he has to be. And, as a seasoned aeronautics engineer and test pilot, he has had a number of determined ideas about the program which have been incorporated into it because Deke held out and proved his point. (He did not win out on the three-axis control system; it was too late. But his ideas on this problem may be incorporated in future systems destined for lunar flight.) He knows the technical details of his job thoroughly—some observers who are familiar with the day-to-day work of the Astronauts say he is the best engineering test pilot in the group—and he combines this knowledge with an ability to analyze a complex problem and articulate his ideas about it that has made him an invaluable member of the team—whether he is ever allowed to fly a mission or not. But Deke does all of this best in small groups. After the press conference in Washington at which the Astronauts were first unveiled, Slayton confided to friends that it was "the worst stress test I've ever been through. If I hadn't already passed my physicals, they would probably have flunked me right then because my knees were knocking."

Scott Carpenter, who served as John Glenn's backup pilot and was then assigned to take over Slayton's flight a short time later, has a leaner and lither look. Graceful as a gazelle, he is an expert swimmer and acrobat and the acknowledged diving-board virtuoso of the group. He is also an excellent dancer—he was voted the best dancer in high school—and though he is bashful about doing a dance like the Twist in public because he feels it might not conform to the proper image of the Astronauts, Scott enjoys the ebullient beat of the music so much and finds the energetic gyrations so relaxing after a hard day's work that he is an accomplished and enthusiastic devotee.

Scott Carpenter is an intense, pensive and sensitive young man. He is also extremely candid about himself. He has admitted that he was a reckless waster of time in his youth. He nearly killed himself in a souped-up jalopy; he flunked out of the University of Colorado twice (because, ironically enough, he could not master the intricacies of heat transfer) and he was awarded his degree only after he had made his own orbital flight and returned home for a hero's welcome. Scott had a conscience *about all this, however, and when he first joined Project Mercury, his friends detected signs of a slight inferiority complex which they attributed to his relative lack of big-league flying experience. Even then, however, Scott's precise and rather high-pitched voice articulated, with immense intensity and eloquence, the almost boyish thrill and pride that he obviously felt at being an Astronaut.*

Scott's chief interest in life, aside from being a good Astronaut, is his handsome and vivacious family—his wife, Rene, who has followed his work in every detail, and the two handsome boys and two pretty girls who are the apple of his eye. Throughout his training, Scott kept a log in which he recorded his daily activities and sent this on to Rene so she would understand exactly what he was up to. An extremely articulate man, Scott filled the log with the same crisp and often poetic imagery which marks his everyday conversation. "Just finished our first rides on the 'wheel,' " he wrote Rene after a session on the centrifuge, "and feel I did some good work . . . The 16-G rides today after three days' practice were easier than the first day's 9-G rides, and no dizziness afterwards. The body—what an incredible machine. I enjoy all this—every minute of it and all facets of it—and wish I could cut my sleeping in half and enjoy that much more . . . If this comes to a fatal, screaming end for me, I will have three main regrets: I will have lost the chance to contribute to my children's preparation for life on this planet; I will miss the pleasure of making love to you when you are a grandmother, and I will never have learned to play the guitar well."

Wally Schirra happens to be the only one of the seven who does not wear his hair either close-cropped or in a crew cut. He is also perhaps the most naturally jovial and outgoing man on the team. Wally is genuinely fond of people, and he likes to talk, earnestly and for as long as his listener can absorb it all, about the technical intricacies of space flight and the complex theories of aviation in general which obviously fascinate him. But no matter what the subject is, Schirra also likes to leave his audience laughing. He returned from a tour of duty at the Mercury tracking station in Australia chock full of stories about the good humor of the Australians. "They have a special code expression for 'O.K.' in their checkouts out there," he explained. "It seems to help break the tension. The expression is 'extra grouse.' The first time they tried it on me," he went on, "I thought I was following the wrong sport. Somebody said, 'Site status green, flight control is extra grouse.' 'What in blazes is extra grouse?' I asked. 'It's the same as "Ichi-ban," ' they said. 'Ah, so,' I said, in my best Japanese accent. I thought the tracking station in Hawaii had a nice touch, too, when I talked to them once. 'This is Hawaiian Cap Com,' the man said, 'Aloha.' "

Like all of his colleagues, Wally is an unrelenting competitor. He carries a small, hard rubber hand grip with him and squeezes it from time to time, first with one hand and then with the other, while he flexes his arm muscles at the same time. "You'd be surprised," he says, "how this sort of thing can build up a man's grip. A strong wrist will come in handy when I'm fighting gravity in a capsule and trying to move the control stick at the same time." Like Glenn, Schirra has to be careful about his weight. "The other guys can eat like pigs," he says, "and it doesn't show up much. I really have to cut down, especially on snacks. If I don't, I can hardly squeeze into my space suit." Schirra also cut out smoking for more than a year when the program began. (Only Carpenter, Slayton, Shepard and Schirra were smokers when they entered Project Mercury. All four quit the habit for a time when they began their training, then took it up again later on. Cigarettes no longer appealed to Slayton, but he liked a good cigar after dinner—"one of life's little luxuries." Shepard bummed a cigarette soon after he reached Grand Bahama Island for the debriefing sessions that followed his Redstone flight. Carpenter went back to a pack a day as he started to help John Glenn prepare for his flight; he felt that he was gaining a little weight around his middle and that smoking kept him from eating too much.) Schirra went back to less than a pack a day. "I'll probably stop again," he said, "if I'm scheduled for one of those long, eighteen-orbit missions. I'm afraid if I don't quit altogether, the urge to light up will be too strong for my nerves."

Like Alan Shepard, Wally Schirra is a bug about sports cars. Shepard had his white Corvette. Soon after Project Mercury began, Schirra bought himself a super-speedy Austin Healy—a bright yellow one.

The youngster of the group is Gordon Cooper. Gordo, as his friends call him, was born in Shawnee, Oklahoma, where his father, a flyer in World War I, served as a district judge. His brown hair, cut short, is *slightly wavy. He speaks with a slow, drawling Oklahoma accent that is slightly reminiscent of comedian Bob Burns, and he has a twangy, sardonic sense of humor to match the accent. Cooper, an Air Force major, is the only Astronaut who owns and flies his own airplane. His attractive wife, Trudy, is the only Astronaut's wife who is a pilot in her own right. Cooper has another love in addition to flying. It is the mountains of Colorado, where his widowed mother lives on a small ranch near Carbondale that she and her son own together. Cooper has his eye, he says, on an adjacent piece of mountain land. There are some lakes full of trout in the high reaches, and, like most of the Astronauts, Cooper is fond of fishing. He came back from fishing a small pond near the launching pads at Canaveral one day and described a strange sound he had heard. "It went grrruuuump, grrruuuump," he said. "It must have been the biggest bullfrog in the whole county." The other Astronauts quickly advised Gordo that he had probably heard a six-foot alligator that had been seen lurking in the same pond. On another occasion, Cooper and Grissom had waded out into the surf with their fishing rods when they suddenly spied a shark swimming along between them and the shore. "We beat it out of there, but fast," Cooper said later.*

In the meantime, while the Colorado mountains beckon, Cooper has his eye focused on the moon. "Since I am the youngest," he says, "I think maybe I have a better chance to have some real good fun in this program than some of the others. I've known pilots who could keep going until they were over fifty. That would give me a good twenty years to go yet. We are already planning and testing out the spacecraft and techniques we will use to get a long way out there. As for myself, I'm planning on getting to the moon. I think I'll get to Mars."

These were the men. This was the raw material for the great adventure —seven highly motivated individuals, each of whom had found his way into the group under his own power and was determined to make good. Each of the men was, and is, an explosive bundle of personal pride, professional convictions and private idiosyncrasies. They were individualistic even when it came to keeping in shape. Except for a few spirited games of handball which they played early in their training, they seldom engaged in competitive sports at all. Scott Carpenter could be seen spending hours at a time bouncing up and down like an accomplished circus acrobat on a trampolin behind his motel. John Glenn took to the beach for solitary runs—as did Grissom and Shepard before their flights. Glenn would also leave his work occasionally to go out onto the nearby Banana River on water skis and slap himself into the water until he was black and blue, trying to perfect a difficult turn.

"You would not expect an amateur to work at it that hard," a friend of Glenn's remarked one afternoon as he watched from the bank. "But these guys are so damned competitive every one of them has to knock himself out just competing with himself."

Because they were so self-sufficient when it came to athletics, physical fitness was one area of their training in which the Astronauts were left completely on their own. A suggestion was made at NASA headquarters early in the program that perhaps a coach ought to be assigned to the Astronauts to help keep them in shape. But Dr. William Douglas, the Astronauts' physician, did not think this was necessary. "They are all big boys," he said. "They are all well motivated and they know the importance of keeping in shape. They will take care of themselves." Bill Douglas found the Astronauts to be good patients. They probably suffered as many colds as any other group, he explained, but their resistance was high and they could throw the germ off without much loss of time. Douglas did keep his eyes open for general signs of stress and tension and tried to keep the training program organized so that it would not burn them out. "They are strong," he explained, "but they are not perfect physical specimens, by any means. They have most of the normal human complaints. And they are all different."

There was one great common denominator shared by all seven, however, which served to separate them at the same time it united them. And this was the driving ambition of each man to be the first among his teammates to go into space. The Astronauts were questioned by the press about this aspect of the program at fairly regular intervals. They made a determined effort each time the question came up to convince the public —and themselves—that it did not really matter. There would be plenty of "firsts" in the program for everyone, they insisted—the first man on a Redstone, the first man in orbit, the first man to fly around the moon, the first man to land on the moon, etc.

"*The second and third and fourth flights may accomplish far, far more scientifically,*" *John Glenn explained a year before Alan Shepard was assigned his mission,* "*than the first flight does. . . . We feel very strongly that this program is much bigger than the problem of whose name happens to be on the first ticket.*"

A few months later, at another press conference, Glenn answered the same question in a slightly different way. "*I think it's obvious,*" *he admitted,* "*that all of us do want to be first. . . . Any time you build up a group such as this to a big mission, why, of course, we would like to be first. . . . [But] the program won't be lost to each of us if we are not first.*"

It was no use trying to avoid the issue, however. The press kept pushing, and finally, at still another press conference, John had to admit that of course he cared, and that it was normal that he did. "*I think,*" *he said,* "*that anyone who doesn't want to be first doesn't belong in this program.*"

The other six men answered the same question with variations on the same theme.

Alan Shepard: "*Of course, I want to be first. The challenge is there, and I've accepted it. Each one of us is aware that his total performance may have a bearing on his being picked. That's why we're all working so hard—even at things like water skiing and keeping our weight down.*"

Wally Schirra: "*I guess we really think as a team on this one. We're all united on wanting to be first. None of us is in this thing for fame or money. But there is the thrill of doing something that no one else has ever done. The guy who goes first will leave a pretty high watermark.*"

Deke Slayton: "*The odds of six-to-one aren't very good. But I wouldn't be in this program if I didn't want to be first.*"

Gus Grissom: "*Each one of us is working hard for his own chance. Naturally, we all want to be first or we wouldn't be here. But we are all making contributions to the group, too. There is so much to learn in such a short time that we have to share it. Nothing is saved for private use. That's the best kind of competition.*"

Gordon Cooper: "*We'd all like to be first, of course. But we can't all be. This is a matter of teamwork, and the man who goes first will be the front man for all of us. I have a tiny, secret advantage, however, over the others. My work with the Redstone has given me quite a few chances to visit Huntsville and the Chrysler factory where they build the boosters. I know for a fact that the Redstones for flights two and three have my name on them. I etched it in the skins of both boosters myself."*

Scott Carpenter: "I believe that the best qualified man will go first. I feel that the choice will be made fairly, so it doesn't really matter to me. I would like to go first, of course. But it's more important to me to be one of the seven Astronauts than to be first. If I am not chosen, I will have had honor enough just being one of this group."

This was all academic, however, when it was time for a mission and the pilot had been chosen. As Alan Shepard soared overhead on the pioneering flight, and as John Glenn took off on our first orbital flight—missions that each of the others had naturally hoped to fly himself—all six closed ranks and worked in harness to make it a success. Glenn's flight was only one more step in a long and tortuous process of trial and error, testing and retesting, plans made and remade. By the time Friendship 7 got off the pad, two other men—named Gagarin and Titov—had already performed similar or even more spectacular feats. And just as this book was going to press, another two named Nikolayev and Popovich carried man's exploration into space even further. But John Glenn's mission was a much more daring and honest gamble. For Glenn was the representative of a free and open society, and he took his chances in full view of the world. Then, when he returned to earth, he was fully prepared—as the first Cosmonauts were not—to share his adventure down to the last detail and to relate the complete story of how the flight had gone, what the bad moments were like, what he saw and how he felt. The same was true after Scott Carpenter's epic mission and, in fact, throughout all stages of the program, when Cooper, Grissom, Schirra, Shepard and Slayton had jobs to do. These were the true pioneers in every sense. For, like all good pioneers, they were anxious to mark the trail well for those who would come after them and to explain what they had done and seen to all those who could not come along. This, their book, is proof of that.

JOHN DILLE for Life
New York, 1962

WE SEVEN

A PAST

John H. Glenn, Jr.

TO DRAW ON

It was quite a moment as the heat started to build up outside Friendship 7. We were in the re-entry phase of the flight now. We had trained for this and knew most of the things that would happen. There was the bright orange glow around the capsule which we had expected. There was the dead silence over the radio as the communications were blacked out as a result of ionization caused by the heat. There was the build-up of G forces again as I was pressed back into the couch. There were also some events that were not normal. I could hear chunks of loose material bumping against the capsule behind me before they went flying by the window. I thought from this that the heatshield might be tearing apart, and this concerned me.

I can imagine what people were thinking down on the ground at this moment. I found out later that even men who were close to the program were worried about the outcome and were sort of holding their breath. It worked out fine, however, because we had spent three years preparing for this moment and had perfected a capsule that was in good shape to bring me through. I was in good shape, too, because we had gone over all the contingencies in our training and had complete confidence that we would pull through no matter what happened.

That aspect of confidence, in fact, is one of the main themes of this book. Project Mercury was a careful test of two big propositions: First, that we were on the right track as we tried to put together a system that could take man into space and bring him home safely. Second, that man himself not only could undertake such a flight but that he was a necessary component of the system. The flight of Friendship 7 proved both these points. On the whole, the system worked well and did its job. Where it failed, the pilot was able to insert his own judgment and skills into the system and make it work for him. The flight would undoubtedly have been a failure if a man had not been aboard to assume control and bring the capsule back. We backed each other up, and we showed that man and machine, working together, have a future in space.

It was a moving and gratifying experience for me to receive the warm welcome of the President, the Vice President and the Congress when I got home and to see the faces and smiles of the millions of people who lined the streets. I tried to point out at the time that I was the representative of a lot of other fine people—including the six other authors of this book—who had done so much to make the flight possible. I also tried to point out that it was not just an adventure we'd been on, that we hadn't just gone up for a ride to prove that Americans could tame rockets and fling test pilots around the earth a few times. If that were the only purpose, it would be a little like saddling up a bunch of knights and telling them to ride off into the dusk with their swords without giving them a mission. We Astronauts have a mission, an extremely serious and important one. We are helping to break the bonds that have kept the human race pinned to the earth so that man can finally begin to explore our space environment, which has such an effect on all of our lives—and perhaps even learn to master it. We live under a thick atmosphere which protects us from the hostile elements of space, but also shields some important secrets from us. We cannot really study some of the stars, not even with the biggest telescopes, because they flicker out and have portions of their light bands obscured by the atmosphere. We can't really count the meteorites that rain down towards us until we set up stations out in space where the meteorites are thickest. We can't measure with real accuracy the deadly radiation that saturates our atmosphere until we have fairly permanent installations, manned by trained scientists, anchored out in space where the rays are bumping into each other and shooting off new ones. We cannot really predict the weather in earnest until we get meteorological stations out in space to spot the storms before they form. Once we have done that, we can alert the farmers weeks in advance, and the crops they save ought to be worth a good share of the cost of our ventures in space.

Some of this work, of course, can be done with instruments. Only man himself, however, has the imagination, curiosity and flexibility to notice the smaller facts and take advantage of the unexpected things that crop up. That is why man is needed in space. It has been our mission, as Astronauts, to help put him there. So far, we have only scratched the surface. We have tested out the systems that can carry man through space and protect him while he is there. We have begun to find out, under carefully controlled conditions, what a man can do in this new environment and what effects, if any, a prolonged journey in space may have on his system. From all this, we hope to plan future expeditions using improved systems and more and more men until we are on our way to what we all feel is one of the most important and worth-while goals that this century promises—a fuller knowledge of man's environment.

We have no doubts at all that this will be accomplished. It will be done step by step, just as we accomplished the first breakthrough with Project Mercury. We also have no illusions about the difficulties involved. We have encountered enough problems already to know that it is a stupendous undertaking. We are also human enough to know that it is an exciting and daring venture. There is no doubt about that.

There is nothing superhuman, however, about being an Astronaut. There is nothing spooky or supernatural about flying in space. I have talked to people, both before and after my orbital flight, who seemed to think that both of these propositions were true and that an Astronaut must have to be some sort of Yogi and put himself into a trance of some kind to go through such an experience. Orbiting the earth at a speed of 17,500 miles an hour—or five miles a second—was far different, of course, from anything I had ever done. The fastest I had flown as a Marine test pilot was 1,100 miles an hour. The highest I had flown was to about 65,000 feet. The apogee or highest point of the capsule's orbit was over eight times that altitude. I saw sights and had experiences and sensations in space that neither I nor any other American had ever seen or experienced before. (See "The Mission.")

If there is one thought, however, that I hope the seven of us get across in this book, it is that space flight, like any other kind of flying, is simply the product of normal human skill and technical proficiency. Both of these have to be of the highest order, but there is nothing mysterious or esoteric about either one of them. Like our country in general, all of us had a lot to learn before we could attempt our first manned flights into space. The engineers and technicians had to start from scratch to develop the capsule and some of the other pieces of hardware which we are using. We had to train ourselves to cope with situations we had never tackled before. We have utilized some technical tools, like the control systems in the capsule, for example, and the world-wide tracking range, which are brand new. But there are two basic factors involved in the success of Project Mercury which are not new. The first is that the program is the result of a tremendous amount of patience and hard work on the part of all the people concerned. There is nothing unique or novel about that. These are proven ingredients in any kind of endeavor. The second factor is that the people who conceived and carried out the program are basically no different from any other group of men except in the particular combination of skills and training which they possess. This has been true of everyone from the directors of Project Mercury down to the most junior member of the launch crew. Their skills, of course, are rather specialized, and most of these men are also among the most dedicated and conscientious people I have ever known. To this extent, perhaps, they are unusual. The point I want to make, however, is that this talent was available when the time came. It did not pop into view as if by magic. It grew out of this nation's experience and progress in other fields; and there is more talent where these men came from. From the standpoint of man himself, then, we have the capacity in this country to go as far and as fast in our manned space programs as we wish to go. Some of the feats that we plan may sound slightly Buck Rogersish—like rendezvousing two separate spacecraft in the midst of space to prepare for a flight to the moon or to other planets far beyond. The men who will man these flights, however, will be as unlike Buck Rogers as the Wright Brothers were unlike Icarus. Whether the crews who go to Mars include one of the original seven Astronauts or not, they will be men *like* Scott Carpenter or Deke Slayton or Alan Shepard or any of the rest of us. They will come from the same kind of background and have the same kind of training that we have had.

The seven of us have been privileged to serve as the pioneers of this program. We have helped to work out some of the kinks in the hardware and to try out training procedures which will be used, with changes here and there, for years to come. We have attracted considerable publicity because of the fact that we were first and were called on to jump the first hurdles. Our personal lives have been rather thoroughly reported. This was unavoidable and, to a certain extent, necessary. For along with the privilege of being first we have also had the responsibility of trying to blaze a trail for those who will come after us and for helping to inform the country about some of the problems which are involved, both human and technical.

Many of the technical matters will be discussed at length in later chapters of this book. In the meantime, in order to help prove the point that Astronauts can and do grow up anywhere, we shall begin our story by outlining our own backgrounds and by describing how we got into the Mercury program in the first place. The reader will probably notice that the seven of us approached the program somewhat differently and that we varied in the details of training and background which we brought to it. This was all to the good. Project Mercury was not looking for septuplets when it started its search for the first seven Astronauts. It was hunting for seven trained, experienced men with ideas and convictions that would be useful in setting up a brand-new program.

I grew up in the small Ohio college town of New Concord. My father ran a local plumbing and heating concern and also owned the Chevrolet agency for a time. When I was sixteen he took me out on the sales lot and made me a present of a rather beat-up old coupé that he had sitting there. This made me the chief chauffeur for my gang of friends—which included Annie Castor, a girl I had known ever since we were both small children, and who has been the rock in my own family for the last nineteen years. The old car, which we called "The Cruiser," also gave me a chance to follow a bent I have always had for mechanical things. I kept it in shape myself. I guess the only time I ever needed help was one day when a group of us took the Cruiser out on a field that had been flooded recently. It was a little muddy, and we had fun sliding around through the weeds. There was a log hidden in the weeds, though, which we couldn't see, and we spun into it and bent the oil pan. We didn't know we had lost all the oil until we drove out onto the highway and burned up the bearings. My father would not let me drive for a month after that, and I learned my lesson.

In 1939, the year of the New York World's Fair, John Glenn borrowed his father's car and drove east with three other boys who had just graduated from New Concord High School to take in the sights of New England, New York City, Gettysburg and Washington, D. C. John did most of the driving. According to Lloyd White, a friend of his who went along and who is now a Presbyterian minister assigned to the Wesleyan Foundation of Ohio State University, "Johnny was the only driver I ever sat behind who enabled me to relax enough to curl up and go to sleep. He was a careful, courteous driver, never reckless. As far as I know, he never even scratched a fender."

I did not think much about space as I was growing up. Later on, of course, when I had been a flyer for a number of years and space flight was emerging as a new frontier of aviation, I began to read up on the subject and to ask a lot of questions about it. As a boy in New Concord, however, I cannot even say that I was crazy about flying. I had the same interest in airplanes that any boy of my age had in those days. I was fascinated from a mechanical standpoint about what made them fly. And our family seldom went for a ride that my cousin and I did not take along little metal airplanes which we stuck out the car window to watch the propellers twirl. One winter, when there was quite an epidemic of scarlet fever in New Concord and I had to spend two or three weeks confined to the house, I got started making model airplanes out of balsa wood. I soon had a room full of these and liked to fly them in the back yard. Occasionally one would crash, and I would repair it and fly it again. In 1938, my junior year at the New Concord High School, I was president of the class. One of my responsibilities was planning the program for the Junior-Senior Banquet. Perhaps I was more interested in flying than I realized, because I picked aviation as the theme. On the cover of the program, which we had printed in red and blue, was a picture of a Ford trimotor airplane, which had been an early symbol of flying, along with the inscription: "NCHS Senior Airways, Inc. Dependable Safe Service. Port NCHS to Port Brown Chapel." Brown Chapel was one of the main buildings up the hill at Muskingum College, where many of us went to school after we graduated from NCHS.

Muskingum is a Presbyterian liberal arts college. I told the professor who interviewed me before my freshman year began that I was interested primarily in technical work. My major was chemical engineering. I was still not interested in aviation as a profession. This interest developed rather quickly, however. I was a junior at Muskingum when World War II began. Prior to the war, the government had set up programs of civilian pilot training to let young men start learning how to fly while they were completing their education. One of these schools was established at New Philadelphia, Ohio, about 48 miles from New Concord. Daily transportation was provided to New Philadelphia, so four of my friends and I decided to learn how to fly. My parents were concerned at first. They thought that flying was still a little too dangerous. Dr. Paul Martin, who was our physics professor at Muskingum, came over to our home one day to discuss it with them. He said that he was convinced aviation had a wonderful future, and that in a few years it would probably be one of the largest and most important industries in the country. My parents and I talked it over again and they finally agreed. The college provided our group with a brown Chevrolet station wagon which we used twice a week to get to classes. We would drive over in the early afternoon, attend classes or fly until it was dark, then drive home again at night. It was my responsibility to take care of the station wagon, and I tried to keep it in as good shape as I had the old Cruiser. Needless to say, I did not take it out into any wet fields.

I was sold on flying as soon as I had a taste of it. As a matter of fact, I think I would have tried to get into some branch of flying along about this time even if the war had not come along. My friends and I each had approximately 65 hours of flying time in the course. We were solo for about 30 of these hours. I was one of the first in the group to go up alone, and we were all very excited about it. This was in 1941, just before Pearl Harbor.

My friends and I took the Army Air Corps physical examination. We passed it, signed up and were sworn in. Then we went home to wait for our orders. But none came. We finally got tired of waiting and went to the Navy and took another physical exam. Once more, we passed it, signed up and were sworn in. This time the orders came right away and we left. It's just possible that I've been AWOL from the Army ever since. I went to the University of Iowa for my preflight training. I went on to Olathe, Kansas, for primary training and finished up with advanced training at Corpus Christi, Texas. While I was there I learned that I could volunteer for duty in the Marine Corps and receive my commission in the Marines instead of the Navy. I had always admired the Marine Corps, so I decided to apply. I won my wings and lieutenant's bars in 1943 and went home to New Concord just long enough to marry Annie.

After some advanced training, I joined a Marine fighter squadron in the Pacific and spent a year flying 57 combat missions in the vicinity of the Marshall Islands. Then I came home to help train other pilots and do some test work at Patuxent River, Maryland, putting brand-new airplanes through simulated combat missions. The war ended while I was in the U.S. I went to a training unit on the West Coast and then had a tour

of patrol duty off China and in Guam. After that I came home again and
spent almost three years as a flight instructor at Corpus Christi. After that
I was stationed for two years at Quantico, Virginia, where I attended a
Marine Corps course in amphibious warfare.

The Korean War had broken out meanwhile, and I went back to combat. I flew F9F Panther jets for 63 ground-support missions, some of
which were along the front and some of which went deeper into enemy
territory. The other pilots in the squadron began calling me "Ol' Magnet
Tail" because I was hit by antiaircraft on seven occasions. Later, I was
assigned to fly with the Air Force as a Marine exchange pilot. I was in an
F-86 fighter-interceptor squadron now, and its main job was to keep
MIGs from coming across the Yalu when there were other activities going on. I had wanted to get a chance at air-to-air jet combat even before
I arrived in Korea. My F-86 flying was just getting under way in the
closing days of the war. In the last nine days of combat I was able to shoot
down three MIGs.

When my Korean tour was completed, I applied for duty at the Navy's
Test Pilot School at Patuxent River, and was accepted. At Patuxent, I
helped test most of the Navy's new jets, the fighters in particular, and
was later assigned to duty in Washington, with the Fighter Design
Branch of the Navy's Bureau of Aeronautics. This gave me a chance to
gain experience in the design of new planes and equipment and to get a
pilot's viewpoint of many new design items. In July, 1957, I flew an F8U
which set a transcontinental speed record of 3 hours, 23 minutes and 8.4
seconds from Los Angeles to New York.

Although I was not conscious of it at the time, all of these various duties prepared me for the advent of the space age and Project Mercury.
The same qualifications that are necessary for a good military test pilot—
the ability to work with large, high-powered engines and new control
techniques and to react quickly and coolly in a difficult situation—are also
basic requirements for a space pilot. In both professions, you must be
able to analyze your own situation rationally and take appropriate action
almost by instinct.

The same thing is true of combat. When you are hit by antiaircraft
fire, for example, you have to assess everything about your airplane immediately and decide whether you can keep it in the air. Are you on fire?
you ask yourself. Has your engine still got power? Have you been hit in
an oil line or a fuel line? If so, are you leaking fuel and oil so fast that you
won't be able to make it back to your own lines? Do your controls still

work and can you manage the airplane? You test them to find out. Most important of all, no matter what is wrong, are you likely to blow up or can you possibly glide back to a safe area without bailing out and losing your aircraft? Decisions must often be made in seconds. I have learned since that heatshield and retro-package problems, re-entry times and thruster problems can also create situations that require rapid decisions.

One day over Korea, for example, I got hit by antiaircraft fire just as I was starting on an enemy target with a load of napalm. At an altitude of 8,000 feet, as I was rolling in to make the attack, I heard and felt a big bang outside the plane, and it rolled sharply to the left. The pilot behind me radioed that he saw a big ball of smoke and that I had been hit. I still had control and turned immediately towards the sea, about 30 miles away, figuring that if I had to ditch I would rather wind up in the water, which our Navy pretty well controlled, rather than in Communist territory. The enemy shell had knocked out part of my trim controls, along with a three-foot section of the right wing. I was able to get up to about 15,000 feet, where I slowed to landing speed and determined that I could still control the airplane in that condition. I headed for our base near Pohang, and made it home with no further difficulty. After I landed, we saw that the shell—which was about the equivalent of our own 90-millimeter antiaircraft—had hit the napalm tank first, under the wing. This had partially cushioned the explosion and saved the airplane from being torn apart. Though we found pieces of shrapnel as big as a man's thumb inside the fuel tank, the tank had sealed itself up around the holes and had left me enough fuel to get home. I still have some of those fragments.

On another mission in Korea, I spotted an antiaircraft gun firing at us as we completed a low-altitude bombing run. I went back after the gun position—a procedure which was not exactly recommended. This gunner evidently had a friend across the valley whom I had not seen, and just as I made a run on him the unseen friend opened up and hit the tail of my airplane. The F9F that I was flying nosed over as it lost trim control. This happened at a very low altitude, right on the treetops, and it was the closest I ever came to crashing in my life. I was down low because I wanted to knock out that gunner so badly that I kept firing at him until I had to pull out just to keep from hitting him with my own airplane. The air speed was about 400 knots at the time, and the only thing that saved me was that I reacted by training and instinct. I jerked back on the stick as soon as the nose started to dip and skimmed right over the top of a rice paddy—almost mushing into it. Fortunately, the controls worked.

I had only partial power until I got back to about 10,000 feet; but then everything started working again, and I went on home. There was a hole in the tail big enough to stick your head through. We figured I was hit this time by a 37-millimeter shell. If it had been a 90, it would probably have knocked the whole tail right off.

A similar test of training and reaction came on another occasion when I was flying F-86 Sabre jets along the Yalu. We had been briefed to drop down over North Korea and go after targets of opportunity on days when the MIGs failed to come out. On one of these occasions, we were strafing troops and trucks that we had spotted moving down a road about 30 miles south of the Yalu. My squadron commander, Lieutenant Colonel John Giraudo, was just ahead of me and had finished his run when an antiaircraft shell knocked out his controls. John tried to make it to the ocean, five or six miles away, but his plane was completely out of control and it was behaving now like a helpless derelict in the sky. It made un-controlled dives and climbs until John, who was afraid the next dive would crash him into a mountain, bailed out. I followed him down and circled his position for quite a while, hoping to keep enemy troops away from him until a rescue helicopter could arrive. I wanted to stay but was running low on fuel. I estimated the fuel it would take to permit me to climb to altitude and cruise to the point where I could glide back over our own lines and either make a dead-stick landing or bail out. I stayed until I had just that much fuel left before I climbed up to altitude, about 40,000 feet, and headed for home. The engine flamed out as planned and there was a strong unexpected tailwind at altitude which helped my glide range considerably. I made it back to our home field and completed the only honest-to-goodness dead-stick landing I ever had to make. I ran into the operations building just long enough to pinpoint on the maps the place where John had landed in his parachute, and took off again with three other planes that the operations officer had ready for us. We could not spot John, however. We found out later that the North Koreans had captured him. He came back during the prisoner exchange when the war was over. I was at Freedom Village the day he was released, and we had a fine reunion.

I do not tell these war stories because I recommend that young space pilots ought to go through such experiences. You do all kinds of things in wartime that you would not normally do. But I do believe that events like this in our background did help to instill confidence in us that we could handle ourselves and our equipment in a crisis. You go out into an unknown every time you fly in combat, and you often face stress situations that are far more exacting than the mere physical strain of pulling Gs. Through such experiences, and by constant training, combat pilots build up the experience required for quick reaction which they can rely on, almost without thinking, whenever they get into trouble.

Test piloting is also good experience. The stresses are not quite the same because you do not have someone shooting at you. But by working with the designers and engineers on a brand-new, complicated airplane, you learn to ferret out the bugs and problems before they can be built into the system to worry other pilots who will use later production aircraft. At Patuxent, for example, I was doing some low-altitude, maximum high-speed work with a new airplane and firing the guns at the same time under simulated combat conditions. Aerodynamic forces on the airplane were very high during these maneuvers. During one of the long runs, a big section of the right wing tip broke off. This took away quite a chunk of the lifting surface, and the airplane rolled slightly. I still seemed to have control. I slowed immediately and went through a standard routine to determine exactly how much control I had at various speeds and whether it was adequate to make a safe landing. Even if you can maintain control of an airplane at its cruising speed, there is always the question of whether you can maintain the same control when you slow down to land, because the flow characteristics change. I tried flying first with the gear and flaps up, then with the gear and flaps down to see if the characteristics of the airplane changed appreciably. I found I did not have as good control over the airplane at landing speed as I would have liked. Turbulence along the ground might also cause some loss of lift, and I certainly didn't want to be caught in the midst of a stall if something happened at the last minute. Following normal procedures in such a case, I came in with extra air speed, and touched down in good shape. Once more, it was the kind of reaction that you build with training and experience. Even if a plane got completely out of control, I do not think an experienced pilot would panic. There would be no point in that. He would still be sitting there trying to figure out what to do right up to the last minute or until he had to bail out.

I began to learn more and more about space while I was on duty at Patuxent and in Washington. I read everything I could find on the subject and kept my ears open. Bob Moore, an officer who sat at the desk next to mine in Washington, was responsible for some of the Navy's liaison with the X-15 program, which was making preliminary studies of man's ability to manipulate a winged vehicle in space. I was very interested in the information that he obtained. Then our office was asked to furnish a test pilot to visit the NASA Laboratory at Langley Air Force Base in Virginia and make some runs on one of the space-flight simulators as part of a NASA investigation of various re-entry shapes. The same officer was supposed to go on from Langley to the Naval Air Development Center at Johnsville, Pennsylvania. He would make runs on the large centrifuge there in order to compare data obtained from the simulator with data obtained from the centrifuge while under high G forces. I requested and was given this assignment, and, although I had no inkling at the time that I might turn out to be one of the Astronauts myself, I enjoyed this stint very much. I spent a few days at Langley and more than a week at Johnsville, where I helped work out a mission on the centrifuge that simulated the conditions a pilot would get into as he made a re-entry from space. It was an interesting taste of the future.

Then another matter cropped up. NASA requested service participation in drawing up plans for the mock-up of the Mercury capsule, which it was already considering, pending the selection of the Astronauts, who would have something to say about it themselves. Since I had been to Johnsville and knew something about the ride the capsule would have to take, and had also been on a number of mock-up boards in the Navy and was familiar with the procedures, my boss in Bu Aer, Captain Tony Benjes, arranged for me to go to the McDonnell plant in St. Louis, where the capsule mock-up was being discussed, and act as one of the service advisors to the mock-up board.

It occurred to me along about this time that perhaps I ought to try getting into this program myself. It began to sound very interesting. I knew that I might have a couple of small strikes against me. Thanks to the timing of the war, I was not a college graduate—although service schools and additional university work I had completed while I was in service had given me the equivalent of a college degree. Also, at thirty-seven, I was probably a little older than most of the men NASA was considering. I was philosophical about my chances, however. I was a senior major in the Marine Corps, and if I were promoted to lieutenant colonel I would probably be in line to command a Marine air squadron. My career, then, was in pretty good shape. I did know that space travel was at the frontier of my profession, and I naturally wanted to be in on it. But the gratification and simple pride over the fact that I might have enough brains and stamina and experience to be chosen was the least important of my reasons for wanting to try. I felt that many of my experiences had added up to prepare me for the kind of challenge that Project Mercury presented and that I would be remiss if I did not volunteer to put some of this background to good use.

When the screening of military test pilots was completed, my name was on the list of those who met the minimum requirements, along with over 100 others. This list was pared down through several screenings until finally I was one of 32 prospective Astronauts—along with Carpenter, Cooper, Grissom, Schirra, Shepard, Slayton and 25 other pilots. NASA asked us to take a series of tests which would help weed us out further. Some of these were physical tests, to measure exactly how much stamina each of us had. Some of the tests were psychological, to measure our maturity and alertness and see what motivated us. As one of the doctors who was in charge of thinking up some of these psychological tests put it, "An Astronaut is not going to be a large mouse, cat, dog or monkey placed in an apparatus which carries him off the earth. He is a unique kind of creature who has emotional needs and reactions which play an important role in his general behavior and in the way he will approach the tasks necessary on a space flight."

I must have passed on that score. From some of the strange questions the doctors asked us, it was hard to imagine what they were really looking for. We joked later, after we had been through the final selection, that perhaps the doctors picked us and let the "good" ones go. There were about 600 questions on the personality-inventory tests alone. Here the doctors picked our brains apart to find out what kind of people we were. They were apparently analyzing our motives for wanting to join the project and were trying to uncover anyone who seemed too shallow to understand the implications of what we were getting into or too egocentric to work together on a team. In one of these exercises, which the doctors called a "Who Am I?" test, we were asked to complete the sentence "I am . . ." twenty times. Each answer supposedly gave the doctors a clue to our personalities. The first few answers were easy: "I am a man; I am a Marine; I am a flyer; I am a husband; I am an officer; I am a father." When you got down near the end it was not so easy to figure out much further just *who* you were.

The medical doctors did not miss anything, either. I am sure we were given the most complete going-over to which any group of men had ever been subjected. We were submerged in tanks of water to measure the displacement of our weight and to determine how much fat there was in our systems. We were given electrocardiograms to check our hearts and elec-troencephalographs to determine certain brain-impulse patterns. There were seventeen different eye tests and several studies of our blood. We had to demonstrate how well we could undergo all kinds of stress and dis-comfort and then snap back again. We walked on treadmills, stepped re-peatedly on and off a 20-inch step, and rode a wheel-less bicycle on which the brake was tightened at intervals to see how we reacted. The doctors kept track of our blood pressures and heart rates while we did all this, and they usually stopped us when our pulses reached about 180. Then, while our hearts slowed down, they checked us again to see how quickly we could come back to normal.

The doctors also placed us in a heat chamber and baked us at 135 de-grees for two hours. They dunked our feet into buckets of ice water, partly to see how we reacted to the shock and partly to measure our blood-pressure response to the cold. We made pressure-suit runs in altitude chambers up to a simulated altitude of 65,000 feet. We all spent several hours sitting in the silence of a dark, soundproofed isolation room to find out how well we could put up with all that quiet. All through this period the doctors, engineers and psychiatrists kept quizzing us about our back-grounds. I did well on most of the physical tests, and was very proud of that.

We had a waiting period of ten or twelve days between our final test and the word on whether or not we had made it. I went back to my desk at BuAer in Washington, and I was there when I got a phone call from Mr. Charles Donlan at NASA asking me if I was still interested. I said, "Yes, I am, very much," and Mr. Donlan said I had made it. I was very proud—I could not help that—but I also felt a certain humility. I could not help that, either—not when I saw all of the scientific talent which was being poured into the program and when I realized how important it was to the nation that we succeed. That night Annie and I had a dual cele-bration. It was also our wedding anniversary. We went out for dinner and took in a play.

Scott Carpenter carries the quiver and gives marksmanship lessons to his two sons, Scotty and Jay. His wife, Rene, leans over to assist one of the two Carpenter daughters.

The Astronauts are all family men, and on this and the next two pages they are shown taking time out to relax with their wives and children. John Glenn, who kept himself in shape by running on the beach at Cape Canaveral, is joined in a workout by his wife, Annie, daughter, Lyn, and son, David.

Wally Schirra, a sports car enthusiast, starts out for a spin in his Austin Healey with his wife, Jo, daughter, Suzanne, and son, Marty.

Deke Slayton joins his wife, Marjorie, and son, Kent, at a motel pool near Cape Canaveral where the Astronauts and their families often gathered during vacations.

Al Shepard and his wife, Louise, who have two daughters, play a family twosome shortly after his flight on the first manned Redstone.

Gus Grissom gets the whole family together—wife, Betty, sons Scott and Mark, and the dog—for a portrait.

Gordon Cooper, his wife, Trudy, and their daughters, Janita and Camala, park the family airplane after a flight. Trudy is the only licensed pilot among the wives.

SIX TO FILL

Malcolm Scott Carpenter

Leroy Gordon Cooper, Jr.

Virgil I. Grissom

Walter M. Schirra, Jr.

Alan B. Shepard, Jr.

Donald K. Slayton

THE TEAM

In this chapter, John Glenn's colleagues describe how each of them happened to join Project Mercury and some of the background which led up to their decisions. Their accounts are arranged in alphabetical order, starting with Commander Carpenter.

A CHANCE FOR IMMORTALITY

Malcolm Scott Carpenter

As JOHN GLENN HAS SAID, all seven of us had slightly different reasons for wanting to become Astronauts. I suppose, in all honesty, that I had the most to gain of any of us from the standpoint of my career. This was not my only consideration by any means. I was motivated by some very strong ideals, too. But in 1959, when the invitations were going out for test pilots to come and take a look at Project Mercury and see if they were interested in joining up, I was not too happy with what I was doing and I was eager to make a change.

Like most of us, I got into flying as a result of the war. As a boy in Colorado, where my mother and I lived, I was much more interested in an athletic, outdoor life than in anything else. I think I really wanted to be a rancher. But the war started while I was still going to high school in Boulder, and after I graduated I signed up for the Navy pilot training program at Colorado College. I was too young to see action. But I went through a series of preflight and flight training programs at various schools in California and Iowa before I wound up back in Boulder again, where I attended the University of Colorado and started working towards a degree in aeronautical engineering. With this training under my belt, I joined the Navy and became a flyer. After more training in flying and electronics, I was sent to Hawaii and put to work flying missions on big multiengine P2V patrol planes. These aircraft are chock-full of electronics and communications gear and spend most of their time flying over water on shipping surveillance. This was exactly what I did during the Korean War, as a matter of fact, while some of the other fellows were shooting down MIGs or plastering the front line with napalm.

In 1954 I received orders to the Navy Test Pilot School at Patuxent River, Maryland. I had a chance there to fly every kind of operational and experimental airplane in the Navy stable. Some of this was antenna work—flying around in boring patterns and circles to find out how well the radio antennas worked on various aircraft at different altitudes—but it was a fascinating time for me, probably the highpoint of my career until then. Then the Navy shifted me into intelligence work. This turned out to be good experience. I learned quite a bit about photo reconnaissance and received some more training in navigation as well as electronics and communications. I think all of this helped me some when NASA looked me over for Project Mercury. The people at NASA were looking for pilots with different kinds of talent so that the Astronauts on the team would complement each other. They apparently needed someone who understood a little more about communications and navigation than the average pilot would have learned, and I could help out here. I still felt slightly self-conscious about my flying, however. Most of the other fellows I was competing with had a total of more than one or even two thousand hours in jets. I had only 2,800 hours of any kind of flying, and only 300 of this was jet time.

I suppose this should not have concerned me, but it did. I was worried that I had not had a real chance to prove myself. My nerves and reactions were all right, however, and my record was good. One night I happened to be up on a training flight with a student pilot when we lost an engine. The poor man was so scared that he could hardly talk when I told him to call the tower and tell them his problem. But I was elated with myself when I saw that I was not the least bit shaky. I knew the plane well, and went through the normal procedures of feathering the engine and getting rid of the drag. There was plenty of power left in the other engines, and we made it. That was all there was to it. It just took practice, and I had lots of that.

I also had a little problem the first time I took a plane out on a night flight as a plane commander. We were flying out of Barber's Point in Hawaii, and we were supposed to go about 250 miles out on a certain leg, then fly back to get a navigational check, then go out again. It was about an eight-hour mission and the night was very black. I had finished the first two legs and the next one took us out over a really dark sea. There were thunderstorms in the area, and I could see flashes of lightning playing around. I knew there was going to be some turbulence, and it was not exactly pleasant to turn my tail on the friendly lights of Honolulu and keep on going. Looking back on it now, it sounds a bit silly. But it takes little moments like that to build up a person's tolerance of fear and his ability to face the unknown.

I had two other chances to prove myself to myself when I was a passenger on transport planes. Both cases involved a hatch that had come open. The first time was on a Navy plane when the entrance hatch blew wide open. This put such a strong drag on one side of the aircraft that the pilot thought for a moment he had lost an engine. The crew chief, who was standing by the hatch, tried to close it by himself. He stood there tugging at it with no parachute or safety strap on, and I was afraid he was going to fall out. I wondered why no one got up to help him, so I did. On the other occasion, I was riding in a NASA plane when an emergency exit hatch blew wide open. I remember thinking, Are we going too fast for you to stick your arm out and pull that hatch back in against the stream? I guess the answer to my question was no, because I just reached out and pulled it shut. I didn't expect a medal or anything on either of these two occasions. But, still, I was rather pleased with myself. I was cool, and I was happy to see that I was.

I had another problem worrying me, however, when Project Mercury came along. I had been tapped for sea duty. Actually, this is the kind of thing that every naval officer has to go through, whether he's a flyer or a sailor. But in this case I was assigned to serve as an intelligence officer on an aircraft carrier, and I really did not want to go. It would take me away from my family for long stretches at a time. And it would take me away from flying. I would be confined on the carrier to a desk job. This was a pretty bleak prospect for a man who wanted to fly as badly as I did.

I was already at sea when a letter arrived from the Navy Bureau of Personnel in Washington. The letter went something like this: "You will soon receive orders to OP-05 in Washington in connection with a special project. Please do not discuss the matter with anyone or speculate on the purpose of the orders, as any prior identification of yourself with the project might prejudice that project." That was all I knew. I was intrigued, but I spoke to no one about it. The next day, when the mail plane came in, the orders were on it. Since I was now in the intelligence game, everyone guessed that I had been called to Washington for some kind of intelligence briefing. I thought so myself.

They flew me off the ship, and when I reached land I called my wife, Rene, and then went home by train and showed her the orders. Neither of us could imagine what this hush-hush project could be. But the next day, when I went to the Los Angeles airport to board a plane for Washington, I picked up a copy of *Time* Magazine and read that 110 pilots with certain qualifications were being ordered to the Pentagon to see if they were qualified for manned space flights. Good Lord! I thought, That couldn't possibly be me! Some friends of mine met me at the airport in Washington, and when we discussed the *Time* piece they asked me what I would decide to do if this *was* it."There's only one decision I can make," I said.

The group I was in got a briefing from Admiral Pirie on what the project was all about. Then we were asked to comment on whether or not we were interested in hearing more. I said I was and I called Rene that night. She had read the *Time* article by now, and all she said when she came on the phone was, "Well, is it or isn't it?" I said, "You're right. What do you think?" Rene answered that she was sold on it. That night I sat down and wrote to her. I explained that I would never do anything foolhardy, but that this was something I very much wanted to do. Then I went through the first phase of tests involved in the selection process. This took a week. We were told to return to our duty stations and that we would be advised if we were chosen to continue.

I had to go to sea again. Each day on the ship, I expected the letter to arrive telling me how I had made out. But when a whole week went by with no answer I figured they had notified everybody who had made the grade and were trying to draw up a real smooth explanation for those of us who hadn't. It turned out, however, that a letter from NASA was sent to my home in Long Beach by mistake. Rene got it on a Tuesday. It said, in effect, "You have been chosen to proceed with further interviews and tests in connection with Project Mercury. Please call this number in Washington by noon on Monday if you wish to continue." This meant that my answer was already a day late. So Rene got on the telephone and informed Washington that I wished to continue. I went back to the mainland two more times to complete the tests which finally weeded thirty-two of us down to the final seven. While I was waiting to see if I had been picked for the team, I had to go back to my ship once more. She was in port at Long Beach now, but she was due to sail almost immediately for another six-month cruise, and I was in for a fairly close call.

The day before the ship left, I got word from the officer of the day to call someone in Newport News. It was Mr. Charles Donlan, the associate director of Project Mercury who was in charge of putting together the team. Mr. Donlan said, "We'd like to have you aboard." I said, "Fine, I'll be your hardest worker." Then I went back to the ship. I was elated. But there was no time to celebrate. Everyone else was carrying his gear aboard while I was carrying mine off. The ship was almost ready to sail. I went in to see the skipper and told him that I expected written orders at any minute transferring me to NASA and that I had already received confirmation of my appointment by telephone. I said I thought I ought to be getting off. The skipper couldn't believe it. He said I could not leave the ship without official orders and that even if I received them he would let me leave only if I promised to be back on the ship in time for maneuvers. It was clear that the captain had not gotten the word. He had no idea that I had been assigned to an important program which would keep me busy for at least three years. I got back on the telephone and told the people at NASA that my ship was leaving in sixteen hours and that if they really wanted me in Project Mercury they had better hurry up and get me off. NASA promised that someone would call my skipper that night and straighten him out. Meanwhile, however, I thought it would be prudent for me to go ashore, get my clothes and bring them *back* to the ship—just in case. The call came through, however, and I had to spend the next few hours trying to straighten out the mess that I was leaving behind in the intelligence section. We were just leaving for a very important operational readiness inspection; the captain had no replacement for me; and on top of all this the intelligence office was in the process of being moved to another part of the ship and was in a state of complete confusion. I think that when I walked off the ship everyone thought I was crazy. I remember that the poor captain asked me what this was all about, and I tried to tell him.

"I am going to ride the nose cone of an Atlas three times around the world," I said.

And he said, "WOW!"

I think one of the main reasons I made the team—aside from the knowledge of navigation and communications that I could bring to it—was that I managed to do quite well on the physical tests. I was the sort of physical specimen they were looking for, I guess, and I was in good shape. The doctors confided to me when they were totting up the scores that I had broken five of their records. The treadmill was one of these. This was a walk that moved uphill underneath you at a constant speed, and you had to keep pace with it with your legs. It was like climbing a mountain that kept getting steeper and steeper, and it was tiring, to say the least. But it was immensely satisfying to me to know that I kept going longer than anyone in my group.

These were the most difficult tests, the ones that tired you out. The doctors had us all wired up, of course, and were listening to all of the important functions of our bodies so they could tell if we were in trouble. We could have *been* in real trouble, no matter how well motivated we were, if we were not in good shape. I also did well on the respiration test which was given to us at the Lovelace Clinic in Albuquerque, New Mexico. The object here was to breathe in room air and then exhale it into rubber bags at the same time you were riding a bicycle that got progressively harder to pedal. The purpose of the test was to trap the air we exhaled in order to determine how efficiently our lungs worked. The doctors had seventeen rubber bags standing by for me to fill. But I fooled them. I kept going so long that they ran out of bags. The doctors said this was the first time that had ever happened, so I think I broke that record, too.

When I finished a test in the altitude chamber, the attendant told me that he had never seen anyone spend two hours at a simulated altitude of 65,000 feet without showing some increase in his pulse rate or his blood pressure. I had just managed to do that, so I felt pretty good. And when I hopped on and off a 20-inch step without letting my pulse shoot up too much, the doctor said, "Boy, you've been wrecking records all over the place, and I think this is another one." He checked and, sure enough, it was. Another record fell when the doctors asked me to blow into a tube that held a column of mercury. The purpose of this was to see how long I could keep blowing before I ran out of wind. The record up until then was 94 seconds. I thought to myself, I'll just count up to one hundred as fast as I think seconds ought to go and I'll try to break this record, too. I counted too slowly, however, and when I finally had to give up I found out I had blown into the tube for 171 seconds without taking a breath. This little victory pleased me very much. The doctors were jumping up and down. I was next to the last in our group to take this test. Then John Glenn came along and managed to blow into the tube for 151 seconds. John did well all through these tests, too. He told me one day that he had overheard the doctors say the two of us were doing so well they were going to tell Washington about us. The funny thing about the breathing test was that a man can store up enough air in one breath to last him for at least seven minutes if he needs it. When you feel you are running out of air and *have* to take another breath, it is not because your oxygen supply is running low—you have oxygen to burn—but simply because the little carbon-dioxide detectors in your system are warning you that the CO_2 concentration in your lungs is piling up above normal and that you ought to clear it out with another breath. All you have to do, if you are taking a test like this and really want to beat it, is to ignore those little CO_2 detectors and keep going. You won't faint. It is just a matter of how much control you have over your own system. Anyone can do it if he is motivated. And I was motivated.

I guess I was motivated on the psychological tests, too. At least, I enjoyed taking them. I especially liked the Rorschach ink-blot test. This is where the psychologists show you blobs of blotted ink and ask you to describe what the patterns look like. Rene had studied abnormal psychology at school and had told me about these tests. She said all she could usually see in the ink blots was a bunch of bats. But she warned me that this sort of upset the psychologists who gave the tests because they had a theory that there was something wrong with anyone who kept seeing bats, one after another. So I went into the test watching out for bats. I saw only a few of them, but I identified them when I did. "That's a bat," I said, the first time a psychologist showed me a blot of ink that looked exactly like a black creature with wings on it. He rose up out of his chair with a look of fascination and asked, "Which way is it flying?" "Straight at us," I said. I don't know whether this was the right answer, but he let me go. We also had to arrange some wooden blocks the way children do in a nursery. This seemed kind of silly, but I assumed the doctors knew what it was they were looking for, so I did not mind.

I did pretty well on the "Idiot Box." This was an electronic gadget full of flashing lights and buzzing buzzers which was supposed to rattle us. You were allowed five seconds to straighten out the mess, and you were supposed to push buttons and pull levers and do all kinds of things in combination to keep the buzzers quiet and make the lights go out. A recorder kept score on you. The box got so frantic sometimes that if you sat behind someone else while he was banging away at it the effect was really hilarious. The doctors ran the box first at normal speed for 30 minutes. Then we had to manage it at double speed for 35 minutes. Finally, the doctors bore down and speeded the thing up to quadruple speed for the last 40 minutes. The man who made the contraption claimed that no one could keep up with it even at normal speed. But some of us managed to stay on top of all the lights and buzzers at quadruple speed, and the people at Wright-Patterson Air Force Base, where we took the test, were amazed. I guess they thought we would panic. They even had the soles of our feet instrumented, because skin resistance is a good measurement of whether or not a subject is panicking. We got so wrapped up trying to beat the box, however, that the work went very smoothly. I think the final portion of this test was probably the fastest 40 minutes that I ever spent. It was a good test of coordination and fast reflex action—the same kind of qualities that we would be needing when we went up in the capsule.

I could not understand the value of the "Who Am I?" test. All you had to do was think of a long list of words which described you, and this seemed so simple that I could not imagine why we had to take it. I answered, "I am a man. I am a naval aviator. I am a father. I am a husband. I am thirty-three. I am an archer. I am a swimmer. I am a skindiver," etc. I filled out all of the blanks, and in the very last one I wrote, "I am a prospective Mercury Astronaut."

Perhaps that answer helped, too. At any rate, I got on the team and I got off the ship. The next thing I had to do, before I settled down at the Mercury headquarters at Langley Air Force Base in Virginia, was move my family there from California. I was getting things ready to pack at our home in Long Beach when the landlord brought a lady over to show her the house. He had told her who I was. She came into the room where I was working and asked me if I was really one of those new spacemen. I told her I was, and she said, "You're a nut!" Then she walked off to look at the kitchen.

I guess she was not the only one who thought that. My father wrote me from Colorado that he was overwhelmed I was getting the chance to pioneer on such a grand scale. But he said some of his friends were wondering why I would volunteer for such a dangerous and untried mission. I could have told them that I volunteered for a number of reasons. One of these, quite frankly, was that I thought this was a chance for immortality. Pioneering in space was something that I would willingly give my life for. As I told Rene, I think a person is very fortunate to have something in life that he can care that much about.

FLYING IS IN MY BLOOD

Leroy Gordon Cooper, Jr.

I CAN THANK MY DAD for my flying career. I started flying with him when I was a young boy, and flying has been in my blood ever since. Dad never had any formal flight training, but he became an excellent pilot. He was in the Navy during World War I. Then, after the war, he became a member of the Army National Guard. He was out of the service for several years. But soon after World War II was declared, he entered active duty with the Army Air Corps. He retired as an Air Force colonel a few years ago. While he was still in civilian life, and was living in Shawnee, Oklahoma, where I grew up, Dad bought a Commandaire biplane which had been designed as a World War I fighter-trainer. It was such a powerful little bird that it could almost do a loop on take-off.

I used to listen while my dad talked flying with some of the real greats of those days—people like Amelia Earhart and Wiley Post, who were among his flying acquaintances. And I was familiar with various flying maneuvers and how to handle an aircraft long before I was in my teens. I did not take any formal lessons, however, until I was fourteen or fifteen years old. When I signed up with an instructor, Dad told me that this was one luxury he would help me finance. I earned some of the money for my lessons by working on the flight line, washing aircraft and helping out with odd jobs.

It was quite a while, however, before I got into flying as a career. The Army and Navy flying schools were not taking any candidates the year I left high school. So, since I could not pursue a flying career, I decided to enlist in the Marine Corps. I left for the Parris Island boot camp as soon as I graduated. The war ended, however, before I could get into combat. I was assigned then to the Naval Academy Prep School and was an alternate for an appointment to Annapolis. The man who was the primary appointee made the grade, however, so I was reassigned in the Marines and wound up on guard duty in Washington, D.C. I was serving with the Presidential Honor Guard in Washington when I was released from duty along with other Marine reservists.

After my discharge, I went to Hawaii to live with my parents. My dad was assigned to Hickam Field at the time. I started attending the University of Hawaii, and there I met my wife, Trudy. She was quite active in flying and was planning to complete the requirements for an instructor's license so she could teach flying. We were married in Honolulu and lived there for two more years while I continued my studies at the University. We did quite a bit of flying together around the islands.

While I was at the University I received a commission in the U.S. Army ROTC. I transferred to the Air Force and was called to active duty for flight training in the United States. After I received my wings I was assigned to the 525th Fighter Bomber Squadron, which was stationed at Neubiberg Air Force Base in Munich, Germany. I spent almost four years there, flying F-84s and F-86s, and managing to get a few credits at the University of Maryland European Extension by attending night school. When I returned to the U.S. I attended the Air Force Institute of Technology at Wright-Patterson Air Force Base in Dayton, Ohio, for two years. I graduated there with a bachelor's degree in aeronautical engineering and was very pleased when I was sent as a student to the Air Force Experimental Test Pilot School at Edwards Air Force Base in California. When I graduated from the school I was assigned to the Fighter Section of the Flight Test Engineering Division at Edwards. This was an excellent job, and I was fortunate enough to become involved in some extremely interesting test programs.

Then, one day, I read that the McDonnell Aircraft Corporation in St. Louis had been awarded a contract to build a space capsule. This really interested me. Flying has always been one of the most fascinating and satisfying things in the world to me, and now that there appeared to be a possibility that we might start flying in space, I could almost taste it. I had been thinking about space, as a matter of fact, ever since I was a kid. I used to read Buck Rogers in the funny papers and wonder if the world was going to have to wait until the twenty-fifth century before people might actually go out into space—not just for the thrill of it, but to explore and discover new things and find out what the planets were really like. Later on, when I started to fly some pretty high-performance airplanes and was getting up to what we then considered high altitudes and high speeds, I always had what I think is the natural desire of most pilots to want to go a little bit higher and faster. I thought that we ought to be trying to extend man's capacities up there.

I also had the idea that there might be some interesting forms of life out in space for us to discover and get acquainted with. I don't believe in fairy tales, but as far as I am concerned there have been far too many unexplained examples of unidentified objects sighted around this earth for us to rule out the possibility that some form of life exists out beyond our own world. I certainly don't pretend that the examples we know about necessarily prove anything. But the fact that many experienced pilots had reported strange sights which cannot easily be explained did heighten my curiosity about space. Plain curiosity was not a prime reason for my wanting to become an Astronaut, but I do believe that any explorer or researcher must have a certain amount of curiosity in his make-up in order to drive him to discover things which have not yet been found. This was one of several reasons, then, why I wanted to become an Astronaut.

I soon found out that Project Mercury was interested in me, too. A few days after I read the announcement about the new space capsule, I was called to Washington, D.C., for some kind of briefing. The reason for my trip was classified, but I had a hunch that it might have something to do with that capsule.

NASA engineers spent an entire morning giving us a technical rundown on Project Mercury and what the Astronauts' part in it would be. When I saw what a logical step-by-step program NASA had in mind, and what a major role the Astronauts would have in it—not only as pilots, but in the engineering development of the program as well—I was convinced that I wanted to be a part of it.

We were asked later in the day to give our reactions to what we had seen and heard, and to indicate whether or not we were interested. I said that I was definitely sold on the program and that I very much wanted to become an Astronaut. I knew quite a few of the other pilots who were being considered, and I could see that there was going to be some stiff competition for the team. I felt that I would be very lucky if I made the grade, but I wanted a chance to fly a spacecraft so much that I was determined to do my best in the competition—and hope for success.

First, we had to take several series of technical and psychological tests. These took hours, and they were hard work, but it was rather fun, too. Next came the physical examinations at the Lovelace Clinic in Albuquerque, New Mexico. Some of these tests were extremely unpleasant. I think the doctors there found a few places to examine that we didn't even know we had.

Then came the physiological or stress tests at Wright-Patterson. These were real challenges. The doctors at Wright-Pat really knew how to separate the tigers from the pussycats. We were isolated, vibrated, whirled, heated, frozen, fatigued and run to high altitude. Through it all we got batteries of psychological interviews and exams. I felt, when they were all over, that I had done very well.

I had every confidence when I returned to Edwards that I would make the team. I even told my boss that he ought to start looking for a replacement. I alerted my family to be ready to move, and we took two weeks' leave to get ready to pull up stakes. I could not be positive that I would make the grade. But I felt confident, and I intended to be ready. I was not really surprised when the phone call came two days after I had returned from leave. The man on the other end welcomed me to the team and asked me when I could leave for Langley Air Force Base in Virginia, where Project Mercury had its headquarters. "How about right now?" I answered.

PROUD TO HELP OUT

Virgil I. Grissom

MY CAREER WAS IN REAL GOOD SHAPE when Project Mercury came knocking on the door, so it was a big decision for me to make. I figured I had one of the best jobs in the Air Force, and I was working with fine people. I was stationed at the flight-test center at Wright-Patterson, and I was flying a wide variety of airplanes and giving them a lot of different tests.

It was a job that I thoroughly enjoyed. So I had quite a few questions to ask myself. If I returned to the Air Force again after three or four years with Project Mercury, would I have to go back as a green hand? That was one problem. Another was that a lot of people—including me—thought that the project sounded a little too much like a stunt instead of like a serious research program. It looked from a distance as if the man they were searching for was only going to be a passenger. I did not want to be just that. I liked flying too much. And I had never been much of a science-fiction or Buck Rogers fan. I was more interested in what was going on right now than in the centuries to come.

The more I learned about Project Mercury, however, the more I felt I might be able to help. I had been interested in engineering ever since I was a kid—even before I knew what it was. And I figured that I had enough flying experience to be able to handle myself on any kind of shoot-the-chute they wanted to put me on. In fact, I knew darned well I could.

I entered the Air Force as a cadet in 1944, after graduating from high school in Mitchell, Indiana, where my father worked for the Baltimore & Ohio Railroad. I realized soon after I got into the Air Force, however, that I needed more technical training if I was going to get ahead. So I took time out to go to Purdue University and get a degree in mechanical engineering. I was married by now, and while I went to school my wife, Betty, helped support both of us by working as a long-distance telephone operator. As soon as I had my degree, I returned to the Air Force, finished my cadet training and won my wings.

I was sent to Korea soon after that, and this was excellent training for what came later. I was assigned to remain on duty there until I had completed one hundred missions—or got knocked out—whichever came first. We chased the MIGs around, and the MIGs chased us around, and I usually got shot at more than I got to shoot at them. I decided space flight could not be much more dangerous than that. You get used to handling yourself in a situation like this, when death is supposedly knocking at your door. You are scared, but you learn to take care of yourself.

We had a rule in our unit that a pilot had to stand up on the Air Force bus that took us back and forth between the barracks and the flight line until some MIG pilot had shot at him, personally. I got to take a seat after my second mission. And I still had ninety-eight more to go. I usually flew wing position in combat, to protect the flanks of other pilots and keep an eye open for any MIGs that might be coming across. Sometimes a bogey would sneak in and start firing at you before you could spot him. There was no time in a spot like that to *get* scared. You had just enough time to call your flight leader on the radio—right now—and tell him, in a calm voice that wouldn't rattle him, that it was time for him to break away fast and get out of there. You also had to remember to tell him which way to break. And you had to make sure to use the correct call sign so you wouldn't get all of the other planes in the flight breaking away at the same time and ruin the mission. This was a lot to do in a split second, and it was good experience. I never did get hit and neither did any of the leaders that I flew wing for.

When I got back to the States, I served a hitch teaching some younger pilots how to fly. This kind of duty is probably even more dangerous than combat. At least you know what a MIG is going to do. Some of these kids were pretty green and careless sometimes, and you had to think fast and act cool or they could kill both of you. I remember one young cadet whom I had up with me one day in our two-seater trainer. He was supposed to join up on another pilot's wing while I watched him work from the rear seat. You come in fast on a maneuver like this. Then you throttle back and glide in right next to the other plane without over-shooting the other pilot's position or wasting too much time. The cadet I was teaching came in *very* fast, and then he throttled back. But he happened to be wearing an identification bracelet on his wrist, and the darned thing got caught in the handle that controls the flaps. The flaps came down at high speed, but one of them broke off and did not come down all the way. We were off balance, so we started rolling immediately. I did not know right away what had happened. I thought that my crazy cadet was causing the rolling, so I grabbed the controls. But even then, while *I* worked on them, we still kept rolling. I knew from experience that only one or two things could be causing this to happen. So I immediately reached for the flap handle. One flap dropped back into place about half-way, so I had *some* control now, and we managed to get back in one piece. It was the kind of situation in which you don't have time to get scared until you're back on the ground, or maybe all the way back to the barracks. Then all you can do is have a beer and think it over.

After my stint as an instructor I put in some more time at Air Force schools studying aeronautical engineering and experimental test-flying. Then I wound up at Dayton as a test pilot specializing in checking out new fighter planes. When the invitation came for me to go to Washington and I learned what it was about, I was willing to hear what they had to say about the space program and Project Mercury. But I must admit that I was fairly skeptical about whether I really wanted to join up.

Then one of the fellows I worked with at Wright-Patterson got the NASA briefing a week before I was scheduled for it. When he came back to Dayton, he was so enthusiastic about it that I decided there must be something to it. I got the same briefing myself the following week. And I liked the sound of it even more. I figured there was a lot I could learn about flying in this program if I got to go into orbit first—or even if I got to go last. And I decided that maybe I could help the program along a bit. I did not think that my chances were very big when I saw some of the other men who were competing for the team. They were a good group, and I had a lot of respect for them. But I decided to give it the old school try and to take some of NASA's tests.

The only one that I did not do too well on was the treadmill. My heartbeat went up to 200 beats per minute before the doctors stopped the test. I was real disappointed in myself, and I thought I should have done better. I did pretty well, I think, with the "Idiot Box." This was the instrument panel full of lights and buttons and levers that we had to push and pull. I guess they had taken everything they could think of that had ever gone wrong on an Air Force plane and put it onto one panel. Everything was out of phase, the lights were going on and the buzzers were sounding, and they watched you to see how you reacted in a clutch situation like this. They timed you to see how fast you could think and use your fingers without getting flustered. The doctors knew most of this already from our records, but they wanted to see for themselves. I guess they thought we were getting a little older now and might get more flustered. This test was not easy.

The heat chamber, however, was simple. Even though they heated me up to about 135 degrees, I managed to keep cool and skim through an old *Reader's Digest* that I had read a couple of times before, just to keep from getting bored. They had a TV set hooked up in the room where we tried on the pressure suit. There was no sound, but I watched the picture while the test was going on.

The psychological tests were a real bother to me. I tried not to give the headshrinkers anything more than they were actually asking for. At least, I played it cool and tried not to talk myself into a hole. I did not have the slightest idea what they were trying to prove, but I tried to be honest with them and give the doctors straightforward answers without getting carried away and elaborating too much. I was optimistic when all the tests were over with. I had compared notes here and there with some of the other fellows in my group, and I thought I had probably done as well as any of them had.

After I had made the grade, I would lie in bed once in a while at night and think of the capsule and the booster and ask myself, "Now what in hell do you want to get up on that thing for?" I wondered about this especially when I thought about Betty and the two boys. But I knew the answer: We all like to be respected in our fields. I happened to be a career officer in the military—and, I think, a deeply patriotic one. If my country decided that I was one of the better qualified people for this new mission, then I was proud and happy to help out. I guess there was also a spirit of pioneering and adventure involved in the decision. As I told a friend of mine once who asked me why I joined Mercury, I think if I had been alive 150 years ago I might have wanted to go out and help open up the West.

A POLYP IN TIME

Walter M. Schirra, Jr.

MY INTEREST IN FLYING was apparently bred into me. My father, who is a civil engineer now, was an ace in World War I. After the war, when we lived in New Jersey, he used to fly his old Jenny in air circuses now and then just to keep his hand in. My mother went barnstorming with him and she would walk the wings of the plane to help pay for the fuel and the hangar bills.

My father did not push me into flying. But we were very close, and I had such complete respect for him that I wanted to be just like him. Clyde Pangborn was a friend of Dad's. I used to pedal 25 miles to Teterboro Airport from our house in Oradell, New Jersey, just to talk to Pangborn for a while and watch the airplanes. I was flying Dad's plane by the time I was fifteen. I never got to solo in it, however. Dad loaned the plane to a friend of his one day, and the friend pranged the plane into the ground and broke it to pieces. I never got to solo, as a matter of fact, until I joined the Navy and started flight training.

I did not reach this stage, however, until 1947. After I graduated from junior high school in Oradell and the Dwight Morrow High School in Englewood, I spent a year at the Newark College of Engineering, brushing up on mathematics and similar subjects before I plunged into the curriculum at the United States Naval Academy at Annapolis. I graduated in 1945, and one of the first assignments I received was to the U.S. battle cruiser *Alaska*. I also served for a time in China on the staff of the Seventh Fleet. I left the surface Navy in 1946 to commence flight training at last and got my wings in 1948. After three years with a carrier-based fighter squadron, I went to Korea as an exchange pilot on loan from the Navy to the U.S. Air Force. I managed to shoot down two MIGs—though the official score reads that I got one kill and a "probable" kill on the other one. When I returned to the U.S. I went into test-pilot work at the Naval Ordnance Test Station at China Lake, California, where I participated in the development of air-to-air missiles. Then I went back to the fleet for a three-year tour, first in a training squadron and then in an operational squadron, to help introduce the new air-to-air missiles to the fleet. During the operational tour, I became qualified as a night all-weather pilot, which prepared me for a billet at the Naval Air Test Center at Patuxent. I arrived there in January of 1958.

This was the most satisfying work that I could ask for as a pilot. A test pilot is fiercely proud of his profession. He knows that he is helping to develop fine new airplanes to defend his country. When you are a test pilot, you not only have to know your airplane better and fly it better than anyone else can, but you also have to be able to explain it, in complicated engineering terms, to the experts who have designed it and are building it. You have to know all there is to know about how many G forces the plane will sustain and how efficiently it burns its fuel. You have to figure out all of its idiocyncrasies and have all of its standards pinned down to a hair. Then you have to know how to apply these standards. When the plane is ready, *you* have to take the first crack at it before it has a chance to kill some other poor pilot who has to take it into battle and trust it. It's a great responsibility. When I first heard about Mercury, I was just starting work on a brand-new airplane—the F4H. This was the latest thing in Navy aircraft, and I knew it would be a great addition to the fleet. I also knew that when that duty was over I would probably wind up being given command of an F4H squadron and going to sea with it on a carrier. This kind of assignment, of course, is the real ambition of any Navy pilot. So here I was, with a fine career it had taken me thirteen years to build up, plus the promise of a really satisfying future looming ahead of me if I just stuck to my knitting.

I had not thought about space flight at all. I guess the thing that finally decided me—in addition to the fact that, once I got to NASA and heard the briefings, they seemed to make sense—was the same instinct that had inspired me as a test pilot. This was to improve the flying breed and push the frontiers out so the whole nation would benefit. In Korea, for example, I was flying a plane that could *not* go as high as a MIG. I had to go on the defensive every time I went into combat and accept the *enemy's* attack. That did not seem very American to me—at least not day after day. You only have two cheeks to turn, and they get kind of worn out. Well, here we were, starting out something new—space flight. I figured that perhaps there were some new standards to set here, too, and that maybe I could help.

I also felt that I had the experience and the qualifications for the job. I knew there were unknown dangers involved and that the Astronauts would be using a brand-new kind of flight system. But I was certain that my years of combat flying and test piloting had prepared me to handle any kind of emergency that anyone would want to try on me. No pilot ever knows, of course, until he has tried it. You never really know whether you can hack a challenge until you have seen it. The proof is in the crisis. I do know that you cannot always tell from just looking at a person whether he can hack it or not. I remember I had two squadron mates once. They were both close friends of mine. One of them was the typical Jack Armstrong All-American boy. He was cocky as hell. The other was a real Caspar Milquetoast. I had it figured out that Caspar would probably tie up very badly and leave our outfit after two or three missions. But I was wrong. It was Jack Armstrong who fouled up. Caspar turned out to be a real tiger.

Like any athlete, a fighter pilot always believes he is going to *win* until he has been proven wrong. This is the only way he can face his job. You check your guns before you take off and you check your bombs—you know they'll drop off when you want them to. And then, when you do take off and get up in the air, the only thing that prevents you from banging right back into the earth again is this fragile thing put together out of a few strips of aluminum, with a few tougher chunks of metal stuck into the engine to make it go. But you master this thing, and you have confidence in it, and you think it's a great machine. You *know* it will carry you through, and you're real complacent. And then, all of a sudden, something goes wrong; something hits you. I got hit by groundfire in Korea once. The plane began to shake, and I really swabbed out that cockpit as I checked over all of those instruments. Some people freeze at this point—and they die. Others just pump out all the adrenalin they have and mesh in all the gears they can and start figuring out what's wrong with this beautiful machine, and where all that complacency went. Well, in this particular plane that I was flying—it was an F-84E—I finally discovered that my right tip tank had been hit and had curled back over the aileron. This was causing it to flutter and buzz. The fluttering was feeding itself through the whole airplane until the whole works was rocking and rolling and shaking me up. I finally cut down my air speed and managed to shake the tip tank loose and drop it. The shaking stopped. The wing had a whole batch of wrinkles in it when I landed, and I knew it had been a pretty bad situation. There was not much time to think about that, though, when this airplane and I had only each other for company up in the air. At a time like this, a pilot has to be cool and collected.

I had another pretty good test of my nerves after Korea when I was helping start out the Sidewinder missile program at the Navy Ordnance Station at China Lake, California. The Sidewinder is an extremely clever antiaircraft missile which one airplane fires at an enemy airplane to blow it out of the skies. The missile seeks out the engine of the enemy plane, just from the heat of it, and then flies right up the enemy's tail pipe before it explodes. I got to fire the first Sidewinder at a drone target to see what would happen. I was in an F3D night-fighter, and right after I let the Sidewinder loose it went a little haywire and started a loop which would cause it to chase me instead of the drone. Here was something trying to kill me, and I wasn't even mad at it. I was trying to help it along. All I could think of at the time was that I could not let this little jerk climb up *my* tail pipe. So I made a fast loop, trying to stay behind it. I simply wanted to keep its front end from ever seeing my back end. Obviously, I succeeded, although the test engineer who was with me suffered slightly from the "clanks," which is pilot talk for the shakes.

But these were just things that I thought about as I was considering Project Mercury. I really was not keenly interested in being an Astronaut at first. The more I heard about the idea, the more interested I got. But when one of the NASA people asked me after the first briefing session what I thought about it all, I told him quite frankly that I was not about to chuck a thirteen-year career and a fairly promising future in the Navy on the basis of one short briefing. I wanted some more time to think about it.

This seemed fine with NASA, and I went on to take the tests. I was still on the fence during the first few tests. I knew that someone would sneak in on me back at Patuxent while I was gone and take that juicy F4H project away from me. And I wasn't sure how I would make out on the tests, anyway. There were some younger fellows among us, some fine young athletes. And I thought a few of them could probably beat me. But then I gradually began to realize that I was not doing too badly. Putting up with the Gs on the centrifuge was no problem. And I was used to pulling zero G on purpose when I was flying around from place to place in an airplane. Any pilot can do this just by rolling over in a 90-degree bank and then pushing the stick forward slightly to take the Gs off the airplane. Pilots often do this on long, tiresome trips to raise themselves up in their seats and take the pressure off their behinds. Any pilot worth his salt also knows that when he finds himself in certain unusual attitudes—like flying on his back—he can get himself out of trouble by reducing the G load and going towards weightlessness. So this ballistic-type maneuver which we now faced in a rocket was quite similar to one which we had already learned to perform as a routine recovery maneuver in an aircraft.

The altitude chamber was no problem, either. I had practiced in the chamber at Patuxent and had been up to higher simulated heights than they tried on us at Dayton during the tests, so this was easy. I pumped the wheel-less bicycle for fifteen minutes with the doctor cheering me all the way. It was quite an endurance test—they wanted to determine the absolute limits of my pulse and heartbeat—and I was exhausted when it was over. But I did fairly well. The 20-inch Harvard step test was quite tiring, too. You have to go up the step on the stroke of a metronome tick once every two seconds for five minutes. The exertion really adds up. But I knew I had passed it. I finally decided, as I got some of these worst hurdles behind me, that I might as well burn the bridges that I had already crossed and go all-out for a place on the Astronaut team. I knew I would miss that F4H project. But this one was beginning to look challenging, too.

Some of the fellows did not particularly like the tests in which the psychiatrists went to work on them. I guess I just like people. I am always intrigued when people try to delve into other people's minds. So I let the doctors play around with me to their hearts' content. I am afraid, however, that they got too much data on us to ever figure us out. We answered so many questions about ourselves that the doctors will probably be trying to add it all up when we're already on our way back from the moon. There was one test that I was a little dubious about—because I had never gone through anything similar to it before. This was the isolation chamber. The room was pitch-black, with no sound or light. And when you can't hear *anything* in a room, you can imagine that you hear a lot. I beat that test by going to sleep in the middle of it. The doctors had me sit in a chair and they asked me not to get on the bed which was nearby because they were using it for some other tests. But they didn't say anything about not using the pillow that I found on the bed. So I stuck it on the back of the chair and fell sound asleep. They had to wake me up when the test was over.

I ran into only one major snag before the doctors were through with me. This happened at the Lovelace Clinic in Albuquerque when the eye, ear, nose and throat specialist got hold of me. This doctor had been playing around with my ears and nose and now he was looking farther down my throat than anybody had ever looked in my life. He asked me to say "Ahhhhhh" and "Eeeeee," and then he held onto my tongue and started to say "Hmmmmmm." He went to get another doctor and this man "Hmmmmmmd" while *he* held onto my tongue. I started breaking out in a cold sweat. Finally, Dr. Lovelace came in and they told me. I had a polyp, a chunk of swollen membrane, on my larynx. One of the doctors mentioned the word "tumor," and then I really started to sweat. But everyone assured me it was benign, and they offered to take it out right there. I suggested we ought to consult the Navy first, since my body and everything in it still belonged to the admirals. Dr. Lovelace said, "Don't worry. There's no problem. You can take care of this when you go back to Patuxent." The fact that they were willing to work on me themselves, though, was the first glimmer I'd had that maybe I was going to make the team and was being separated from the other boys.

I could not take care of the polyp right away because part of the procedure before they could operate on it was to keep me absolutely quiet for four days and not let me speak. I could not possibly do that just then, because I was due in Dayton to take some psychological tests. I knew that the doctors there would *want* me to talk. Later on, the medics did put me on a week's silent treatment. I had to break it only once when a NASA official called me up from Langley to ask me how my polyp was coming along. I told him he had just interrupted the cure. During most of this time I communicated with my wife, Jo, by writing notes.

Then one day the doctors decided to go ahead and operate immediately. They arranged to give me what was normally a two- or three-month treatment in two or three days to get it over with. That, of course, was the bona fide proof that I was on the bandwagon and that they really cared. The doctor who was lined up to operate on me and remove the polyp was not too pleased over all the rush. He was called in on the double and told to go to work. He looked me over and frowned a few times. Then he said, rather gruffly, "From all the fuss they're making over you, you must be getting ready to go to the moon or something."

He did not know until later what a good guesser he was.

THE URGE TO PIONEER

Alan B. Shepard, Jr.

I GREW UP in East Derry, New Hampshire, where my father, a retired Army colonel, is now an insurance broker. I was raised, if not exactly in an atmosphere of aviation, at least in the midst of mechanical things. I had a five-horsepower outboard motor which I used to take apart and put back together again. And I often helped my father when he had things to tinker with—as you usually do in a small farming town. When I was in high school, a friend of mine and I used to cycle out to the airport, which was about 10 miles from town, and do odd jobs around the hangar in exchange for a chance to take rides in an airplane now and then. I was also keen on building model airplanes. So I was interested in flying at a rather early age.

I graduated from Pinkerton Academy in Derry, then studied for a year at Admiral Farragut Academy in New Jersey to prepare for Annapolis. I graduated from the Naval Academy in 1944. The Navy had a rule that even prospective flyers had to go to sea first, so I spent some time on a destroyer in the Pacific during the closing days of World War II. I took flight training at the Navy schools at Corpus Christi and Pensacola, Florida. Then I served in a fighter squadron that was based out of Norfolk and made two cruises aboard carriers in the Mediterranean during 1948 and 1949.

My flying career really got going in 1950 when I was still a lieutenant, junior grade, and was lucky enough to be chosen to attend the Navy test pilot school at Patuxent River. This was a real plum, especially for a junior officer. Only three of us were picked from my Navy contemporaries. During my two years of duty at Patuxent I helped to perfect such advanced Navy aircraft as the F2H-3 Banshee, F3H Demon, F8U Crusader, F11F Tiger, F4D Skyray and F5D Skylancer. I also did some high-altitude flying to make studies of light and air masses over the North American continent, helped to develop the Navy's in-flight refueling system and was involved in testing out the first angled deck on a U.S. Navy carrier—an idea that we borrowed from the British to make more efficient use of the carrier deck. I was operations officer for a while in a night-fighter squadron that operated off the West Coast and served aboard a carrier in the Pacific from 1953 to 1956.

This last duty, plus the experiences I had checking out new aircraft, pretty well firmed up my reflexes. I was on a night flight off the Korean coast once, intercepting some unidentified planes to make sure who they belonged to. I had been on top of an overcast and was coming back down again into the bad weather. The visibility was restricted, and I got a little off the track trying to find the carrier in the dark. For a few moments I thought I might be in real trouble. "You're running low on gas," I said to myself. "Your radio aids are not working properly. You'll have to ditch in that dark water and take your chances."

Then I realized that this kind of thinking was not getting me anywhere. I knew I would have to settle down and make the best use of the equipment I had that was working, and go through all of the steps that I had learned over the years in my training. By concentrating on the job that had to be done, and by shutting out any trace of fear that was trying to creep in, I got myself back on the track, found my ship and landed.

After my second trip to Patuxent for flight-test work, I was lucky enough to be sent to the Naval War College at Newport to brush up on some academic subjects. After that I became a staff officer at Atlantic Fleet Headquarters in Norfolk, in charge of aircraft readiness for the fleet. This was a fairly diversified build-up. But it was heavily loaded with aviation. And I thought I had a very good chance of becoming the skipper of a carrier squadron in another year or so and going back to sea.

As Wally Schirra has pointed out, running an aircraft squadron is the big objective of any career pilot in the Navy. You are working with young people and helping to contribute directly to the Navy's mission. You are exercising whatever leadership ability you have developed. And you are preparing yourself for even bigger responsibilities of leadership later on. When Project Mercury came along, then, I was fairly well satisfied with my prospects—though I was by no means complacent. Like some of the others, I had quite a debate with myself about what to do.

I had read about the program in the papers before I heard anything about it officially. NASA had published the requirements of the Astronauts it was looking for, and I knew that I was qualified on all counts. I assumed they would probably be looking me up and asking me if I was interested. As it turned out, they were doing just that. But the orders got misplaced somewhere—they wound up on someone else's desk for a few days—and I was beginning to wonder if I had been overlooked or disqualified. The orders finally came asking me to report for the first briefings, and I was delighted. I had a long talk with my wife that night, discussing what I should do if I were selected. Finally Louise said, "Why are you asking me? You know you'll do it, anyway." Louise had always been in complete support of what I had done, and I knew she was behind me now.

I did ask my boss, a Navy captain, for his advice. I wondered how it would affect my Navy career if I took some time out for Project Mercury. The captain told me he thought I had nothing to lose, and that it might be good for the Navy to have me soak up some of these new ideas. Aviation was coming to a crossroads, and space flight was the new turning. It would give me a chance to be not just a pilot but a *space* pilot, and to get in on the ground floor of something new and important.

Without trying to sound too "Navy Blue and Gold," as we say in the service, I really did hope that I had made it. I thought it was definitely a chance to serve my country. And I guess everyone feels an urge to do something no one else has ever done—the urge to pioneer and to accept a challenge and then try to meet it. I realized what it would mean to the nation in prestige and morale. And I felt that I would like to contribute whatever ability and maturity I had achieved. It would also, of course, be a big boost to my own self-confidence to know that I had done well in my chosen field. Every man needs that. But the tests we had to take were not exactly easy, and I was not sure that I had made the team.

I think the toughest tests, in a way, were the psychological ones. I do not happen to be my own favorite subject, actually, and it is always difficult for me to analyze my own feelings or to figure out exactly what is going on in my own brain, and why. There were so many of these kinds of questions—and they always seemed to come out the same no matter how they were worded—that they began to seem repetitious and boring. I do not exactly lack confidence, but it was hard for me to tell them much about myself that they did not already know from my record. When the processing was over, though, I thought I had done pretty well on the physical tests. The examiners played it real tight and tried to keep my group separated so we could not compare notes. And we were encouraged *not* to swap ideas because that would affect the validity of the tests. But we were living pretty closely together, and I thought I could detect from various people's attitudes how well we were doing.

Gordon Cooper and I were in the first group that they looked at, so we had the longest wait of any of the candidates—from the third week in February to April 2—to find out how we'd done. I did hear a grapevine rumor that NASA was not going to pick a team of twelve Astronauts— as we had thought at first—but was narrowing the field down to six or seven. I began to wonder even more about my chances.

It was a Thursday morning when the word came. I was just leaving the office in Norfolk to spend a long weekend in Boston with Louise— we were going to be guests at a family wedding—when the telephone rang. It was Mr. Donlan at NASA. First, he asked me if I was still interested in being an Astronaut. I told him that I was. Then he asked me if I could report to work the following Monday. I said I could, and after I hung up—the office was empty—I let out a loud whoop. I was very happy. I could not reach Louise on the phone, so I drove home without hitting anyone or breaking any laws. Louise and I just held each other after I told her. I could see that she was as happy as I was. Then we flew to Boston, where my parents and sister met us at the airport. As soon as we were all by ourselves I said, "Guess what, Mother—I'm getting out of the Navy." She looked so shocked that I had to hurry up and explain to her that it was just a temporary transfer.

A TEST PILOT'S DREAM

Donald K. Slayton

I GREW UP ON A FARM about five miles from Sparta, Wisconsin, and it was not until I was well into high school that it ever occurred to me I might not follow in my father's footsteps and be a farmer myself. Somehow or other, the idea of flying got into my brain—even though I had never ridden in an airplane and did not even know if I would like it.

Then the war came along and flying seemed like a better idea than ever. I joined the Army Air Corps as an aviation cadet in 1942. Just before I left Wisconsin to begin training, I went out and chartered a five-minute ride in a flying boat that took off from Lake Michigan. I just wanted to see if flying really appealed to me. I enjoyed the ride so much that I decided right then that this would be my life.

I got my wings a year later and was sent to Europe, where I flew fifty-six combat missions in B-25 bombers. I had to pass a swimming test before they would let me go overseas. I had never learned to swim too well. We only had a creek to paddle in near the farm. But they gave us the test in a shallow lake, and I just waded across. I had to do a little better than that, of course, when I became an Astronaut, because we had to learn how to take care of ourselves in the ocean in the event we got dunked during the recovery operation after a space flight.

After my first combat tour I came home and helped train B-25 and A-26 pilots for a year. Then I was shipped off to Okinawa just in time to fly several combat missions over Japan. In 1947 I was back home again and out of the Air Force. I entered the University of Minnesota, and by doubling up in my classes I took the full four-year course and graduated in two years with a degree in aeronautical engineering. The Boeing Aircraft Company hired me as an engineer, and I worked in Seattle for a couple of years. It was sort of a glorified draftsman's job, but I did some work on electrical systems and wing designs. Then, in 1951, I was recalled by the Air National Guard. I had switched from bombers to fighter planes by now, and I went to Germany as a fighter pilot and flew out of the U.S. Air Force Base at Bittburg. I met my wife, Marjorie, while I was in Germany. She was working for the Air Force there, and we had our honeymoon in Paris.

I had no trouble making up my mind about Project Mercury. It hit me just right. Actually, I had the best job in the Air Force when Mercury came along. After Marge and I came back from Germany, I was sent to Edwards Air Force Base in California to become a test pilot. This was the one place in the entire Air Force, I figured, where I could utilize both my engineering background and my piloting experience, and put it all together in what seemed to me the ultimate. So I was not too eager to leave there. But I had been at Edwards for about four years when a new regulation came out limiting us to five years in any one spot. That meant the fun would end soon, anyway. And besides that, I could tell from working on some of our hottest planes that we were just about reaching the limit on what a conventional aircraft could do. You always want to try squeezing a little more out of an airplane. But the drag—the friction of the air in the atmosphere—cuts down the performance of any aircraft. Or the temperatures finally build up so high that when you reach maximum speeds they affect the structure of the plane—they blister the paint on the wing, for example. Pretty soon, you find that you've pushed the state of the aerodynamic art about as far as it will go. I saw this sort of thing closing in, and I had just about settled down to take a two-year course in astronautics and space flight to prepare myself for the *next* step when Project Mercury came along. So it was a real fine coincidence for me.

I had done quite a bit of thinking about space flight. It looked to me like a logical extrapolation from what we had been doing all along with airplanes. Except I knew that when we got into space we would be out of the atmosphere and would have some new problems in controls and navigation and communications to figure out. However, I think when you look back on history you will find that almost all of the big advances like this have been of about the same magnitude. Ten years before Mercury began, Chuck Yeager squeezed through the sound barrier in the X-1 and started a whole new era. At that time there were a lot of people around who shook their heads and said it could not be done. And yet here we were, ten years later, flying around at two and three times that speed—and doing it operationally and thinking nothing of it. I never assumed that switching over to the space age would be much different. We would have a lot of things to learn. But we were right up on the frontier line, and we were ready to learn them.

I was a bit skeptical, however, about whether Project Mercury itself sounded like the best way to begin this new era, and especially whether it could make good use of me. I had heard about MISS ("Man In Space Soonest") an Air Force proposal which had been in the works for approximately two years when Project Mercury came along and supplanted it. From what I heard, I had the impression that this was just a matter of tying a man onto one end of a missile and flinging him out there. I rather doubted that they cared whether they had a trained pilot in there or just any human body. But I decided to go to Washington for the NASA briefings and find out what Project Mercury was all about. I thought perhaps the people in charge could convince me that they really *needed* a pilot for this new program. I decided, however, that I was going to be hard to convince.

As it turned out, it did not take them too long to win me over. As soon as I heard the first briefings that Monday morning I said to myself, "Boy, there's just no one who can hack this thing *but* an experienced pilot." I told the NASA people that I was willing to hear more, and I stayed on for further interviews and got myself into that group of 32 men who took the final tests. I do not normally have hunches, but I figured then that I would be on the list. I even had a feeling that I would be the first Astronaut to go into orbit. I guess I was wrong on that one.

The thing that impressed me most was that they had come to us test pilots instead of to some other group for help. There had been some talk about training scientists how to fly—astronomers, for example, or weather experts—so they would know what to look for when they got up into space. We wound up, however, teaching Astronauts how to be scientific observers. This made a lot more sense. Our main function would be to try out the new systems and help perfect them as we went along. That is what a test pilot is for—to help develop a new system of flight from the ground up and shake the bugs out of it. It is the only profession which is equipped to do this sort of thing. Not just any pilot would do.

In combat, for example, you are thinking about what goes on *outside* of your airplane. You are most concerned about the enemy and the anti-aircraft guns and where your wing man is. If the test pilots have done their job right, you should not have to worry about whether your aircraft is capable of performing its mission. But in test flying you have an entirely different problem. You are concerned about what is going on *inside* the airplane, and what the aircraft itself is doing. This is the dividing line. As a test pilot, you have to evaluate your vehicle, not situations outside of it. We would be doing essentially the same thing in Project Mercury. The man in the capsule would have to be ready to make quick decisions and evaluate systems which were entirely new. No one had ever done this before—at least, no one that *we* could talk to. But the experienced test pilot is trained to run into things that no one has yet written a book about. He never knows what the devil *is* going to happen. He just has to be prepared to cope with it. It is this native ability to realize *when* you are in trouble and to what *degree* you are in trouble that really counts. This is what an experienced test pilot knows almost by instinct. He has to sit there and think, What's going on here and what can I do about it? It occasionally becomes a trial-and-error process. If you try something sensible that won't work, you try something else. But you need experience to know exactly what to try. A green kid—or a nonflyer —could not do it. He might manipulate the wrong controls, panic or jump out too soon. Even a man who is normally a real tiger in combat might not be able to make it—if he did not have the right education to fall back on. If you are an amateur in this business, and you just *think* you are in trouble, you can really *get* yourself into trouble very fast simply by doing the wrong thing first. You might be a whole lot better off if you did nothing at all.

For example, I was helping to test out the F-105 at Edwards, and I was putting the plane through a tight decelerating turn from maximum speed to determine the turning capabilities of the aircraft with limited thrust. Suddenly, as I was winding her down at full power and came subsonic, the aircraft snapped over on me—it did about four or five snap rolls—and I ended up in a flat inverted spin. The F-105 had never been spun before, so I had no information to fall back on. In fact, I was not even sure at first that I was *in* an inverted spin—which means that I was suddenly flying upside down. I decided to try a standard spin-recovery technique that works on some century-series fighters when they are in a normal spin. I applied antispin control and immediately got the distinct feeling that the plan had reversed direction on me. I tried neutralizing the controls. Then I noticed the direction of rotation and tried to counteract it, but all of a sudden the aircraft was going in the opposite direction. After I tried this approach three or four times, it was obvious that the spin was reversing on me faster than I could neutralize the controls. The whole thing started when we were up at about 38,000 feet. I checked the altimeter now and saw that I was going through about the 27,000 mark. We were falling pretty fast.

This particular plane happens to have very effective speed brakes. These are four petal-shaped pieces of metal which you can open up like a flower right around the tail section. I thought, Well, if you pop these, they may act like a kind of spin-chute, put a lot of drag on the aft end of the plane to get your nose down, and then you can work yourself back into a more vertical-type spin that you may have better luck with. I checked the altimeter again about this time and I thought it said 10,000 feet. I thought to myself, Boy, it's getting pretty late. You'd better start thinking about getting out of this mother. I reached down low for the ejection handle to see if I could reach it—this would pop me free of the plane in my parachute if I had to get out. But I was being forced out of the seat and could not reach it. I pulled up on the stick and forced myself back into the seat. I could just barely touch the handle, and I thought, Well, O.K., you can get out of here now if you have to. I was convinced now that I was definitely in an inverted spin because of the forces which were trying to throw me right up into the canopy and clear of the aircraft instead of down into my seat as they would on a normal spin. This meant, as I found out later when I checked the instrumentation, that I was pulling from one to two-and-a-half negative Gs instead of the positive Gs that you usually experience in flying. Throwing out the speed brakes had pulled my nose fairly straight towards the ground and had damped out the pitching oscillations. I tried once more to break out of the spin and noticed a brief hesitation before the direction reversed again. On the next try I felt the spin slow just a bit. I neutralized the controls, and this time it worked. Whammmm! The spinning stopped. We pulled out of the dive, and I went back and landed. That was a fifteen-million-dollar airplane, and I did not want to lose it. But the main point is that I did not have time to worry about that at the time or to be afraid. The whole thing was over in about 40 seconds, and I just had to use my head and try everything I could think of as fast as I could.

Nothing in my training, however, prepared me for some of the tests that we had to take before we could become Astronauts. I was ready for most of them—the ones that determined whether we would be able to stand the pace physically. But I have to admit that a few of the tests were a little over my head. Someone said that we had to answer about 3,000 personal and psychological questions. It seemed to me at that time that there must have been at least five times that many. We spent three whole nights at Dayton just writing out our answers. The headshrinkers showed us some pictures and told us to make up stories to go with them. They even showed me a blank card and asked me to make up a story to go with that. I don't remember what I told them, but this seemed like ridiculous child's play, so I just had some fun with the doctors on that one.

When we were finished I figured that I had made out fairly well—at least on the physiological tests. I pulled 9.2 Gs on the centrifuge before I blacked out. That one was easy for me. We had run out of G-suits while I was testing planes at Edwards, and I spent a year and a half flying in all kinds of conditions without one. That proved to me that you can build up a tolerance for G forces just as you can for anything else. The heat test did not bother me, either. Your face gets kind of warm in that oven, and I had a couple of hot spots on my knees and hands. It was uncomfortable, but you can tolerate it. It was also a pretty good jolt when they stuck our feet into a bucket of ice cubes. But before the seven minutes were up my feet got so cold I couldn't even feel them, so that was not so bad. I got fairly tired on the treadmill at Lovelace. I went until I just couldn't go any more. But it was not really a great stress. I think the Harvard step test we had at Dayton was the hardest. You jump up and down off the platform for about five minutes. Then they throw you on the tilt board, which stands you straight up while they measure your blood pressure and your heart rate. I ran out of steam before I finished on the step. The tilt table didn't bother me. I just relaxed and started to go to sleep. I think this kind of shook them. In spite of later events and the decision of some doctors that I should not fly after all, I think I am in better physical condition today than I was even at the time when I passed these stringent tests and made the team.

I did not have too long to wait after the tests before I got the news that I was in. I was fairly busy at this point anyway, flying back and forth from one air base to another, trying to catch up on my regular work. I found three weeks of paper work piled up on the desk out at Edwards when the tests were over. And I had a string of conferences lined up at Wright-Patterson and at Eglin Air Force Base in Florida to discuss some of the new planes I was involved with. I really had no time during this period to think much at all about becoming an Astronaut. Then Mr. Donlan called me one morning and said, "You've been selected to join us if you are still interested." I said, "Yes, sir. I sure am." And I was. I had always wanted to jump into something no one had ever flown, and go off into an entirely new element where no one had ever been. This was my chance. I was on the ground floor now of something that the human race could be concentrating on for the next thousand years—if we did not destroy ourselves in the meantime.

It was a Friday morning when Mr. Donlan called. He asked me to be in Washington Monday morning. This was a good, fast start.

WE ALL PULL

Donald K. Slayton

Leroy Gordon Cooper, Jr.

Walter M. Schirra, Jr.

TOGETHER

In this chapter, three of the Astronauts team up to describe the concept of teamwork which got Project Mercury going.

A JOB FOR EVERYBODY

Donald K. Slayton

IT WAS OBVIOUS to us from the start that Project Mercury was much too complex and far-reaching for all seven of us to learn everything there was to know about it in every detail. If we had tried to do that—read all the stacks of technical manuals and go out to visit all of the hundreds of contractors and subcontractors who were building bits and pieces of the system—we'd still be back in the classroom. So we split the work up between us. Scott Carpenter, who had done quite a bit of that sort of thing in the Navy, took on communications and navigational aids. Al Shepard, who had experience with ships and Navy headquarters people, concentrated on the tracking range and the recovery teams we would need to pull us out of the water after a flight. John Glenn had been involved in aircraft design, so he bore down on the layout of the instrument panel which we would sit in front of within the capsule. NASA decided we would adopt the U.S. Navy pressure suit for our space suit, so Wally Schirra, a Navy man, started work on the life-support system which we would need to keep the pilot alive and comfortable. Gus Grissom had done a good deal of technical engineering at Wright-Pat, so he worried about the automatic and manual-control systems which we would use to fly the capsule. Gordon Cooper headed out to the Army Redstone Arsenal in Huntsville to check up on the boosters we would use for the ballistic flights. And I pitched in to study the Atlas which we would use for the first orbital missions.

Everything was parceled out fairly evenly. We each went our own way, dug into our subject, and then made reports back to the others whenever we discovered something we thought they ought to know. Scott Carpenter, for example, told us what he was doing about navigational aids. He knew, of course, that the word "navigation" was really a misnomer in our case. In flying, navigation is generally defined as "continuously detecting and correcting infinitesimal errors in the flight path." When you've been sent up by a missile, though, and have been tossed either into a ballistic trajectory or into an orbit, your course is already set. There is nothing you can do to change it. You can control the attitude of your spacecraft—whether it's upside down or backwards or tilted over to one side. And you have to do this to keep it from overheating or tumbling out of control. It is also extremely important to be able to control your attitude when it's time to come home. Otherwise, you may have a rough re-entry.

But changing the attitude has no effect at all on your flight path while you are going around in orbit. Once you have separated from the booster, your course is determined. So the problem of navigation in Mercury boiled down to the simpler matter of knowing exactly where you were at any point along the way. Scott put most of his effort into working on items like the periscope—which we would use to give us a visual presentation of the ground we were covering—the navigational maps we would carry to help pinpoint our location, the various radio and telemetry circuits we would need for communication—which isn't always easy to hang onto when you're soaring around the world 100 miles up at 18,000 miles per hour—and the radar installations and beacons that would keep us spotted so the people on the ground would always know where we were.

Gus Grissom spent so much time worrying about the automatic pilot, which would keep the spacecraft in its proper attitude, that one day one of the officials at NASA turned to him and said, "Gus, if this thing doesn't work, it's all your fault." The official was only kidding, of course. None of us had complete responsibility for his own contribution. After all, there were hundreds of other engineers working full-time on the project, and the seven of us had a lot of other things to do besides helping to design the system. For one thing, we had to start in at the same time and learn how to *use* the darned thing. This meant, in effect, that we had to learn a new way to fly. The attitude control system which the engineers had worked out was a novel kind of gadget for us, and it took quite a bit of time and trouble for us to master it. Basically, the job of keeping the capsule in its correct attitude during flight is taken care of by small jet nozzles fastened on the outside of the spacecraft. These nozzles can squirt out streams of hydrogen-peroxide steam on three different axes, and this action shoves the capsule around to keep it properly aligned or to straighten it out if it gets out of position. If you are rolling from side to side, for example, and you want to stop this motion and get back to an even keel, you activate the jets on either side of the capsule that compensate for rolling. If you happen to be yawing—or twisting along the flight path the way a crab crawls across the beach—you activate another set of jets which push you back into a straight line. And if you're pitching up and down, like the bow of a ship when the waves hit it, you have another bunch of jets mounted outside which can cancel out this motion. you are re-entering the atmosphere, for example, and running into all that Sometimes you may want to *make* the capsule pitch, yaw or roll. When friction and heat, you must get the capsule into a position where its blunt end is pointed straight down into the airflow. You have to *hold* that end in place. And you do it by squirting the jet nozzles which would tend to keep the vehicle properly aligned. The system was designed so that the pilot can handle all this himself. He does it with a single control stick which works all three axes at once—yaw, pitch and roll. When you move the handle back and forth, you activate the two sets of jets outside the capsule which control the pitching—up or down. When you move it sideways, to the right or left, you make the capsule roll to the left or right—or you cancel out a roll that has already started by giving the stick a little push in the opposite direction. When you twist the handle, or rotate it in your hand like a water faucet, you control the yaw axis. It is really not difficult once you get the hang of it. But some of us never did

like the idea of putting all three axes together. Pilots have always controlled the yaw in airplanes by using their feet on the rudders, and most
of us would prefer doing it that way now. But we are stuck with what
we have, so we have all tried to make the best of it. It works. But if you
happen to be pulling a lot of Gs—as we would be during the final portion
of the re-entry back to earth—it might get a little hairy trying to manipulate the controls with all the finesse you'd need. The pilot should be able
to do this. And we have practiced it on the ground until we know we can.
But some of the adjustments the pilot has to worry about have to be very
fine. So we have added the automatic control system—Gus's specialty—to
back up the pilot in case he is incapacitated, or to take over and help out if
it gets to be too difficult on a rough re-entry.

Al Shepard's share of the teamwork broke down into two parts: the
recovery techniques which we would put into play to find and rescue
the Astronaut and his capsule after they had landed, and a complicated
link of radar and communications installations that we set up on ships,
islands and friendly continents around the world to keep an eye on the
spacecraft throughout the trip and transmit any radio commands that
might be necessary to bring it down in an emergency.

Al also took on a few peripheral duties. For example, the parachutes
which lower the spacecraft down into the water after its ride were considered part of the recovery problem. They were designed to scatter out
clouds of radar chaff when they deployed, to help the ships spot the
chutes as they started down. The engineers rigged up an underwater
bomb that would go off when the capsule hit the water to give the Navy
another fix on our location. And Al and his committees had to make sure
that there were enough ships and planes scattered in just the right places
to find the man and fish him out, no matter where he fell, before it could
get dark. We knew there were not enough ships available in the entire
U.S. Navy to plug up every gap. So we had to *predict* the spots ahead of
time, in order to give the ships plenty of time to get out there and get set.
Then we had to hope that the re-entry went according to plan. The
people Al worked with resorted in this case to a branch of probability
mathematics which the scientists call "Monte Carlo"—because it boils
down, really, to nothing more than a complex way of figuring the odds.
They added up everything they knew about the trajectories that our
Atlas boosters had taken in the past and the orbital path the capsule itself
would probably take, and they came up with a number of landing points
that they labeled "high probability" areas. One of these, which extends

out from Cape Canaveral, would catch the spacecraft if there was a failure right after the booster took off. If the Atlas ran out of momentum after that and we just missed going into a good orbit, we knew that the capsule would glide back down in the vicinity of the Canary Islands, just off the west coast of Africa. But if everything went by the book, and the capsule went on around—and the retro-rockets fired right on time to bring it down—we were ready with another landing area located just northeast of Puerto Rico. The Navy stuck most of its recovery ships in these big areas. It also arranged to scatter a few more crews in what we called "light probability" areas, where we *might* have to bring the capsule down in case it had an emergency somewhere along the way and missed the big areas altogether.

Al was also responsible for figuring out the best way to get the Astronaut out of the capsule once he landed. Most flyers have had to learn some kind of ditching technique in case of a forced landing over water, so the system was basically a modification of the standard procedure. The Astronaut would be wearing a fairly watertight pressure suit, and he would have stowed inside the capsule a small inflatable raft which he could toss out if he thought he was going to have a long wait in the water. Attached to the raft he would have a survival kit which would include a mirror he could use to signal airplanes overhead, some packages of shark repellent and a knife for cleaning fish. We tried all of this out in a huge tank at Langley. It had a machine for making waves to make the exercise more realistic, and we had our graduation exercises down at Pensacola, Florida, where we went out on a barge and practiced inflating our rafts in the ocean. Al coordinated this part of our training, so he tried everything out first and showed the rest of us how to do it. He even decided that we ought to work out a scheme for getting out of the capsule from under water. He said he figured this was a possibility we had to face— that one of the capsules might start sinking someday before the Astronaut had time to get his hatch open—and that we ought to worry about it ahead of time so we wouldn't panic. We put a capsule in the tank at Langley, and then Al got inside with skindiving equipment on and had them put the hatch back on and push the capsule under water. He tried to open the hatch and push it aside so he could get out, but the pressure of the water kept banging it back in place. Finally he found that by putting his back against the hatch and shoving real hard with his feet he could push the hatch open against all the pressures. Once the hatch was clear, it would start to sink and he could get out and come on up with no worries. Al explained this underwater idea to the rest of us. We all tried it, and it worked pretty well.

I did the same sort of trouble shooting for the team where the Atlas was concerned. This was my specialty, so I tried to attend all of the co-ordinating meetings which involved figuring out ways to put the capsule on top of the Atlas and mate the two of them together. The capsule was one kind of bird; the Atlas was another. They were being built by different companies hundreds of miles apart, and they were both darned complex and stubborn machines. We assumed there might be something about that capsule sitting on top of it that the Atlas would not like—the extra weight and the aerodynamic roughness, for example—and we had to try and assess this problem ahead of time and make sure that the Atlas people were aware of it and could take it into account as they prepared our boosters. Or there might be something about the Atlas that would not suit the capsule—all that noise and shaking around when the Atlas first ignited, for example—and we had to pass that word along to the capsule people. I monitored both sides for the edification of my colleagues. And since I knew I might sit up there myself some morning and referee those two birds if they weren't getting along, I was probably more interested in what some of the problems were than anyone else in the room. I was performing, essentially, like a test pilot for an aircraft company. I wanted to make sure that the engineers, who were not flyers, were not overlooking anything obvious.

I also watched all of the Atlas tests and launchings at the Cape that I could, and made notes on all of the procedures that might affect our own flights. This turned out to be very valuable. We tried to plot out all the little things that could go wrong at almost any minute in an Atlas count-down—why it went wrong, what you could do about it, and how you could avoid it—and we used this list later to design a fairly foolproof safety system that gave us a loophole out of almost all the different crises we figured we could get into. We discovered, from watching the Atlas perform during launchings and by asking the computers various questions about split-second emergencies that might crop up, that most of the conceivable failures in the Atlas would give us an interval of at least a few seconds between the first sign of trouble and the disaster itself. This period of warning could be dangerously brief, and it might not give us enough time for a man to look everything over himself and react with his own hands. But the Atlas people worked out an ingenious system which they called ASIS—for "Abort Sensing and Implementation System"—which consists chiefly of a series of electronic sensing devices stuck at critical places all over the Atlas which can send a signal to an automatic switch immediately if they detect trouble. These sensing devices measure the fuel pressure inside the Atlas, for example, and the hydraulic pressure in the engines, and keep track of the voltage in the Atlas' electrical system. If the ASIS detects trouble seconds after a launching, it can send a signal to the automatic switch which triggers the escape tower and pulls the capsule free of the Atlas before it has time to blow up or veer off course. Naturally, if we were going to trust a system like that to save our lives, we wanted to know it pretty well. That was part of my job.

I was mostly an interested bystander, however, where the development of ASIS was concerned. It was in good hands with the engineers at Convair Astronautics. But there were a number of other things to consider about the Atlas besides the obvious fact that it might blow up with a man on top of it, and I spent a good deal of time working on these, too. We tried to arrive at the best moment during the countdown, for example, to put the Astronaut inside the capsule. Originally, the technicians planned to let us crawl in about three hours ahead of time. They thought this much time would be required to complete all of the necessary checkouts on the Atlas itself. This would give us an awfully long wait, however, just lying there on the couch twiddling our thumbs and counting butterflies. We were worried that such a long delay might wear the Astronaut out before he ever started the trip, so we looked very hard at the countdown sheets and figured we could start instead at T minus 90 minutes and still get the job done. It takes only five minutes for a man to crawl over the sill of the capsule and get inside. It takes another 10 minutes to fasten all of the oxygen connections, strap the man down and adjust his suit. This brings you down to T minus 75 minutes. Then you have to allow 20 minutes to put the hatch on, bolt it shut and make some pressure checks of the cabin. This brings you and the Atlas out even at T minus 55 minutes on the countdown, and you have just five minutes less than an hour to sit there and wait. Actually, we knew that we could do all of this in even less time if we had to—because I had timed these operations independently. Unfortunately, we have not yet attained this optimum schedule because of various mechanical problems and a conservative checkout procedure.

I also tried to imagine what it would feel like sitting up on top of a live Atlas, and I made a serious study of this moment from a distance. We wanted the first Astronaut who got a chance to ride the thing to be prepared for some of the strange sounds and sensations that he would hear and feel after he got strapped in. Otherwise, some of them might tend to worry him. You hear a lot of mechanical noises at one point, for example, while they are moving the gantry around outside. The capsule vibrates, and the pipes creak a little later on when the technicians start putting the liquid oxygen on board. The whole bird starts to shimmy on you while they test out the gimbaled engines to make sure that the autopilot can move them properly. And finally, near the end, you might hear the first rush of water as they start flooding the pad to keep it cool just before the engines blast on. I also predicted that we would get some weird sensations after taking off. About 30 seconds after staging, for example, the Atlas is supposed to program over into about a 14-degree pitch back towards earth to get you headed into the precise angle for a good orbit. This can be pretty disconcerting if you are not prepared for it—you might take it as an optical illusion and get the idea that you were diving straight back down again. So we told the fellows what to expect and not to be surprised when it happened. We had also noticed in some films of previous launches that when the big booster engines dropped away from the Atlas, the capsule sometimes became enveloped in flames for a second or two. This could be rather frightening if you did not know it was coming and no one had bothered to tell you about it. Actually, the flames beating around you would be a good clue that your engines have staged and that you are really in good shape. In addition to making these studies, I also pitched in and helped to straighten out the escape hatch on the capsule, which had been designed and stuck into the plans before we ever got involved in the program. In the original design—which none of us liked when we first saw it—they planned to have us crawl through the side hatch in the capsule, then close it up and bolt it in such a way that there was no way to get it off again except to unbolt it from the outside. Since the Astronaut could obviously not do this, he was supposed to crawl up through a second hatch on top of the capsule after he had landed in the water. This meant squeezing around the instrument panel and pulling himself up through a short tunnel that was only 20 inches wide. The trouble with this arrangement was that if you went out with your arms over your head, you could not get them back down. If you went out with your arms down, you could not get them up again. Either way, you could easily get stuck. You would also have a difficult time trying to drag your emergency raft along behind you in case you felt you might have to sit in the water for a while. We had quite a few understandable misgivings about this system, and we tossed them to the people at McDonnell Aircraft. We asked them to adapt the entry hatch and convert it into an exit, too. They finally figured out a way to do it. They contrived a system of primer cord which was stuck in place around the edge of the hatch. When the pilot detonated this primer cord by hitting a special plunger, the cord would explode and blow the whole hatch out of his way, bolts and all. Gus Grissom had a little trouble with this gadget, but the design was basically sound. It was a lot better than the hatch we had started out with originally. This one had 70 bolts to lock it in place, and it took a crew of men about 65 minutes just to bolt it on. That was one reason why the Astronaut was scheduled to get in three hours ahead of time, to give the crews plenty of leeway.

John Glenn, whose capsule was picked up by a destroyer while he was still inside, tried to crawl out through the top hatch but found it easier to blow the side hatch and come out that way. Scott Carpenter, remembering that Gus Grissom lost his capsule when his side hatch blew and that John Glenn got quite warm staying inside, elected to go out through the top and take his life raft with him. With perhaps the thinnest hips on the team, Carpenter found it fairly easy. "It was just like crawling out through the top of a bottle," he said.

I timed the technicians one day with the new, explosive hatch, and it took them only 17 minutes to put it on. This saved the Astronaut quite a long wait. I suppose we came fairly close to nit-picking sometimes as we worked some of these details out. But we decided that since we were the test pilots who would be flying this thing, we had a right to stir things up a bit. That was what they had hired us for.

A SHARP KNIFE

Leroy Gordon Cooper, Jr.

I WORKED WITH the Redstone booster just as Deke has with the Atlas. The Redstone was already a well-proven bird when it was first considered for use in Project Mercury. Of course, it had to be made compatible with the Mercury spacecraft, and this took some close coordination and communication between several different agencies. I feel that assigning an Astronaut to help accomplish this paid off, for several reasons. For one thing, I was a military man who had been assigned to a civilian agency, so I could understand the problems on both sides. As an engineer, I could talk the language of the other engineers. And since I was hoping to ride the finished product myself, I could really get immersed in the problems.

Our main concern was to determine whether the Redstone could take the additional weight of the capsule, put up with the increased bending moments and the added complexity, and still meet the reliability requirements for a manned booster. To plan and coordinate all of this took literally hundreds of meetings between representatives of ABMA (the Army Ballistic Missile Agency—later the Marshall Space Flight Center), which had designed the Redstone; the Chrysler Corporation, which was building the Redstone; the McDonnell Aircraft Corporation, which was building the capsule; the NASA Space Task Group; and the various launch organizations at Cape Canaveral. Some of these meetings were extremely interesting. I remember particularly two meetings which adjourned to the homes of Dr. Wernher von Braun and Dr. Jack Kuettner in Huntsville, Alabama, where we sat and drank coffee and talked about space until the wee hours of the morning. Both of these men had given a great deal of serious thought to space since they were young students in Germany, and they were full of fascinating ideas about what can be done in space and what we may find out there.

Like everyone else on the team, I had several development tasks in addition to my own regular training as an Astronaut. One of these was the development of a personal survival knife which we wanted to carry in the capsule with us. We all knew from our experience as pilots that a knife is one of the most valuable aids to survival on both land or water, and we also knew that we would encounter a good deal of both of these elements during our flights. We would be orbiting over oceans and jungle and desert, and we wanted to be prepared for any emergencies. Several manufacturers of knives submitted samples of their products for us to study, but we were not fully satisfied with any of these. I finally decided to contact a man in Orlando, Florida, by the name of Walter ("Bo") Randall, who was well known as an expert designer and manufacturer of special kinds of knives. Randall produced knives for the commandos in World War II, and had written a book on the subject of knife fighting. At the first chance I had I flew to Orlando to see him and discuss our requirements. Bo had never thought much about making knives for space, and he was tremendously enthusiastic about the challenge. He showed us a design which he said could be modified in a number of ways. I had to make several trips back and forth before we settled on a final design, and we exchanged ideas and hardware by mail for quite some time before we made our decision.

The survival knife we wound up with is one of the strongest knives ever made. It is hand-forged and hand-tempered from high-grade Swedish steel and is so sturdy that it can be used like a chisel to cut through steel bolts. You could probably slice your way right through the capsule wall with it if you had to. The handle is a continuation of the blade and is hollowed out to hold several extra survival items like matches, fish hooks, line, etc. Hand grips of hollow micarta are bolted on top of the handle. It is so strong that you can use it for heavy-duty prying without breaking it off. Randall made nine of these knives for us: one for each Astronaut—and two spares. NASA engineers were so impressed with them they had Randall make several more knives, roughly similar to the originals, so we could attach one of these onto the hatch of each capsule as standard equipment for all of the checkout tests.

When Al Shepard made his flight in Freedom 7 he decided to save his personal Randall knife for use later on—he figured it would be more valuable on an orbital flight—and took along one of the knives which Randall had made for NASA. It was lucky that he did, for his mechanical hatch dropped off the capsule while it was being lifted by helicopter to the aircraft carrier, and the knife was lost. Gus Grissom made the same decision. This was lucky, too, for his explosive hatch malfunctioned and blew off, and the entire capsule sank, along with the NASA-Randall knife. John Glenn also took one of the NASA knives along with him on the orbital mission, and brought it back, safe and sound. All seven of us still have our own knives. Randall has also made two more of the special Astronaut knives at the request of the Smithsonian Institution, where they are now on display. He told me that the day he received this request from the Smithsonian was one of the proudest of his life.

Another task of mine has been to serve as chairman of an Emergency Egress Committee which has been responsible for working out procedures for saving the Astronaut in the event of an emergency on the launching pad. Some day, when space travel is more common than it is now, people will take this sort of thing as much for granted as they take firetrucks now at an airport. But the launching of a spacecraft is a tremendously complex operation, and any number of things could go wrong which could result in injury or death to the Astronaut or to the other personnel before he ever left the ground. Whenever you are working with large quantities of highly volatile fuels, explosive pyrotechnics like escape rockets and retro-rockets, unstable chemicals like hydrogen peroxide and complicated pieces of ordnance, you face the possibility of sudden and dangerous emergencies. We had already worked out rescue and recovery procedures for picking up the Astronaut downrange, even in the case of an abort which might bring him down just a few miles out at sea. But we also wanted to cover ourselves in the event of an unforeseen accident right there on the pad.

For example—to take a remote but serious possibility—if the retro-rockets which are mounted under the heatshield of the spacecraft should ignite and explode inadvertently while the spacecraft was still sitting on top of an Atlas on the pad, the upper dome of the thin-skinned Atlas could easily be ruptured. If this happened, the Atlas could collapse and either burn fiercely or explode. The spacecraft could then topple down into the midst of this inferno and be destroyed along with the Atlas.

To prepare for eventualities like this, we have one of the finest fire-fighting crews ever assembled as part of our emergency egress team. The crew is equipped with three M-113 armored personnel carriers which are similar to tanks and which are specially insulated to provide fire and blast protection for the crew. One of the vehicles carries the special pad-rescue squad. The other two are equipped with a very advanced fire-fighting system which would use a new kind of dry chemical that can smother the hottest flame. We also have four large movable nozzles anchored at opposite corners of the launching pad which can be remotely controlled from the blockhouse. The nozzles are capable of pouring tremendous quantities of water or foam to help keep the flames under control while the armored personnel carriers dash in to go to work. The crews are equipped with special fire suits, a fire axe for cutting through the capsule's skin, self-contained breathing devices for each man to prevent him from suffocating in the smoke and fumes, hooks and cables for pulling the spacecraft out of the area, and a pickup device on front of one of the vehicles to carry the spacecraft to safety if it should wind up on the ground enveloped in flames. Each of the carriers is equipped with the latest in communications equipment so every member of the crew is in voice contact and knows exactly what is going on at all times.

In addition to all of these precautions, Pad 14, where the Atlases are launched, is equipped with a tower just 25 feet away from the gantry. We built a special drawbridge on the tower which can be lowered in 30 seconds so that the end of it sits just outside of the Astronaut's hatch. If an emergency arises, the Astronaut can blow off his hatch, scoot across the gangplank and take an express elevator on the tower which would get him to the ground in another 30 seconds. By this time, we could have a personnel carrier on the spot to pick him up and get him out of the area in a minute or so. We also stationed the cherrypicker behind the blockhouse and kept it ready to move up in less than a minute. Now that we had the egress tower, we did not need the cherrypicker as much as we had on the Redstone launches. But we kept it on hand, just in case.

This is just one example of the kind of catastrophe that we all had to prepare for. There are a number of others, and part of our job was to help write out a long list of standard operating procedures which would cover all of the various contingencies and make sure that all of the personnel working on the pad—including the Astronaut—was familiar with them. The rules and procedures are different for each situation that could crop up. If we had a fire in the tail of the booster, at approximately T minus 30 minutes, for example, and had decided that we could not get it under control, the capsule would be aborted immediately and cleared out of the area by the escape tower. The capsule would be lowered to earth by parachute, and we have helicopters and sea-going vehicles standing by during each launch to pick up the Astronaut wherever he lands. If the test conductor decided that the fire could be kept under control, we had another option. We could lower the gangplank from the egress tower, instruct the Astronaut to disarm all switches on his instrument panel, then blow his hatch and run across the gangplank to safety. There were so many variations of things like this that could go wrong at various specific times during the long countdown that we put out a special manual to cover them all and to spell out the detailed procedures for each kind of crisis. It ran to about eighty pages.

I do not want to give the impression that we expect any of these dire emergencies to occur. But safety is one of the main principles in the operation of Project Mercury, and maximum safety can be attained only when you have planned and prepared for the most serious and unforeseen incidents. This is what we did.

It was an uphill battle at times to obtain some of the special equipment that we needed, to recruit the personnel who were trained to use it and establish the standard procedures that all of us felt were necessary. One reason for this was that the first missile-launching requirements were es-

tablished several years ago, before we ever planned to insert a man into
the system. The entire philosophy had to be revised once we got involved
with manned flights. There was also the problem of the time that it took
to get our hands on some of the special items of hardware, and of the red
tape which is often involved in procurement procedures. We felt that we
were in a hurry, however, and there was quite a bit of hustling before the
system was ready for the first manned shots.

Fortunately, we received fine cooperation from all of the agencies concerned. General Leighton I. Davis, who is the commander of the Atlantic
Missile Range, which includes Cape Canaveral, gave us all the help we
asked for. Pan American Airways, which maintains and manages the
range for the Air Force, provided us with excellent men from its own
fire department, its pad safety department and its management level. The
NASA Launch Operations Directorate helped out with the planning and
the building of equipment and published the emergency manuals for us.
McDonnell Aircraft provided us with some of their best technical brains,
as did the NASA Preflight Operations Division. We all got together and
rehearsed our procedures over and over again. And before each flight, the
Astronaut who is scheduled to make it goes out to the pad to familiarize
himself with the equipment and the personnel who will be standing by to
help in the event of an emergency.

It was a very rewarding duty for me. And I believe it helped to have
one of the Astronauts taking a personal interest in this special program.
Some of this may sound like strange work for Astronauts to be doing.
But space travel is not all moon and stars and distant planets. There are
many unusual duties that need to be done before we can launch a man
safely on such a trip.

SOME SÉANCES IN THE ROOM

Walter M. Schirra, Jr.

I WAS VERY RELIEVED that someone as calm and thorough—and stubborn—
as Deke Slayton was worrying about the Atlas. I have often thought,
Suppose I am lying up there on top of that thing, ready to blast off. And
I'm getting a little vibration that I had not expected. The Atlas down be-

low me is banging and whining, and I'm worried that maybe it's going to spoil my day. I can call Slayton on the radio and say, "Hey, Deke, what's all this shaking and rattling I'm getting here?" Deke will have the instruments in front of him in the blockhouse, and he has studied the bird so well that he can come back and say, "I've noticed that myself, Wally. Don't worry about it. That's just the hootenanny valve on the watchamacallit fluttering a little. I saw it do the same thing out at the Convair plant in San Diego. It doesn't mean a thing. Forget it." That's all I would need to know. And maybe I could return the favor some day if Deke was up there and he did not like the way his oxygen system was working. Then it would be my turn to help.

One turn I did for the rest of the team was to run the heat test on the space suit that I had helped perfect for all of us. The suit was a crucial item, and it had to be subjected to all kinds of wear and tear to make sure it would stand the strain of space flight. One of the tests was an uncomfortable few minutes in a heat chamber where we ran the temperature up to a sizzling 180 degrees to see if the suit itself and the oxygen system it would be linked to could take it. To make absolutely sure, we wanted to have an Astronaut sealed up inside the suit so he could see for himself. I made this heat run. There was no need for all of the others to do it. All they needed to know was how the test went and what I thought about it. The other fellows were finding out about other items in the program that I didn't know. So we traded. Al, Gus, Deke and I went out to sea one day, for example, to practice being picked up inside the capsule by a destroyer. There was no sense wasting a whole day for the others. The four of us told John, Scott and Gordo how we made out, and we briefed them on everything we learned. All of us were vitally *interested* in everything, of course. But we did not have the time to *master* everything.

John Glenn, for example, did most of the worrying about the layout of the instrument panel and the cockpit we would ride in. This was what John called a real "opinion area." You can get as many different opinions about the layout of any cockpit as you have pilots who are going to sit in it. And you can probably never satisfy everybody completely. Basically, though, we all felt that some improvements could be made in the original layout. We knew that when we were flying along under the crushing weight of many Gs in the capsule we were not going to have much strength left over for turning our heads back and forth to look at the gauges or moving our hands too far to reach for a handle. We wanted some of the more important gauges put in obvious places so we could see them better. And we wanted some of the buttons and lights grouped by function so we could find them all together instead of scattered all over the place. In general, we asked for a rearrangement of the instrument panel that would bring everything in closer where we could see it and get to it. John rode herd on all this to make sure that all seven of us knew exactly what was being done and that we were all more or less in harmony when the workmen were ready to start building the thing.

Between us, we had a fair amount of prestige around the country, and though we did not always succeed in getting what we wanted, we sometimes ganged up—all seven of us together—and the extra weight helped us to win at least a compromise. We tried not to press our advantage. But whenever we did hit on something that worried all of us, we would lock ourselves up in our office at Langley until we had a solution that satisfied all seven of us. Then we'd go out and go to bat. Sometimes we would have quite an argument among ourselves before we could agree on exactly what it was we were going to fight for. But we usually stayed in our room until we had all of our points straight, so that the other fellows we were about to tackle would not be able to split us apart or find a hole in our argument. We called a session like this a "séance"—because some people thought we were acting like swamis in there, I suppose, and were pulling answers out from under the table. But once we had had a séance, we would really bear down, no matter who was against us—even if it happened to be one of our top bosses. And when the man had heard us out, he would usually have to admit, "Okay, you've had your séance. So you win."

We séanced the new escape hatch for the capsule, for instance, and won our point. We closed ranks with a whole bag of complaints when we made our first trip out to the McDonnell Aircraft plant in St. Louis and got our first look at the capsule that McDonnell was building. The preliminary designs had been roughed out before we were chosen, and the engineers were already putting some mock-up models together when we got there. Right off the bat, we started picking at the design. We spotted some things about the instrument panel that we wanted changed. But the main thing that bothered us was that for some reason the engineers had decided not to provide us with a window so we could look out and see the view. It seems that some engineers just don't think the way a pilot does. It might have been a lot easier—and maybe a little safer—to build a spacecraft with no window in it at all. The engineers did claim that they had tried to design one for us but were afraid the tremendous stresses and heat we would encounter in space might crack it. They also pointed out that they had already stuck on a periscope and a couple of small portholes for us to look through. But that just wasn't good enough. We all felt that a pilot ought to have a clear visual reference to his surroundings, no matter what kind of a craft he's flying. Otherwise he would have trouble keeping his bearings and maneuvering with real efficiency. None of us wanted to die of claustrophobia out there in space, and none of us could see any point in going to all the trouble to get out there in the first place if we were going to be half blind. We were persistent, and we finally got our way. The engineers built us a window. It was not ready in time for Al Shepard's flight, but Gus's capsule had it, and so did the other capsules after that. We may have been seven loners, but we knew how to act as a team when we had to. And we undoubtedly made some of the engineers a little sore at us sometimes with our bright ideas. But we figured that since *we* had to fly the capsule, it ought to be something that *we* wanted, not just something that satisfied the slide-rule pilots. We had all been working test pilots long enough to know that the engineering fraternity was capable of designing an aircraft which was perfect so far as the theories involved were concerned, but which no pilot could possibly fly. So when we did agree to a little compromise, we tried to make darned sure it was in our favor.

We were not a bunch of stiff-necked dogmatists, however. Once, when we made a trip out to McDonnell and got the distinct impression that the poor engineers were holding their breaths for fear we'd come up with something outlandish for them to worry about, we decided to poke a little fun at our own seriousness. We told the engineers that we wanted them to make some provision in the capsule for airsickness. Then we got aboard a commercial airliner and went home, leaving the engineers to brood over this new request. On the plane, John Glenn and Al Shepard decided to follow through on the gag when they spied one of those paper "burp bags" sticking out of the pocket of the seat in front of them. They pulled the bag out and began scribbling on it. "Here is the answer," they wrote, "to the airsickness problem. Please subject this specimen to environmental justification testing of salt spray, fungus, high and low temperatures, vibration, shock and capacity."

When we got home, the bag was mailed to a McDonnell vice president. Nothing came of the project, of course. Everyone knew that we weren't really concerned about airsickness—or, in this case, spacesickness. And even if we were, it was obvious that we would probably not have time to open up our helmets to use burp bags, anyway—fungus or no fungus.

We also held a séance over whether we would continue with more Redstone flights after Al and Gus had made theirs. All seven of us were against flying more ballistic missions, and we had some of the NASA engineers on our side. They agreed with us that we should move right on to the Atlas and start putting someone into orbit. Our bosses were being extra cautious, however, and they had some different ideas. Of course, their responsibility was different. But we held fast. We insisted that we had learned everything we needed to know that the Redstone could teach us. And we argued that it was not necessary for all seven of us to fly a Redstone before we switched over to the Atlas. We felt that would be a dilution of our real effort—which was to go into orbit. The point was finally won. Sometimes the managers at NASA tried to break up our solid front by taking one or two of us at a time into their offices and trying to swing us over. But once we had séanced something, we usually stuck together. That doesn't mean, however, that we always won.

But though we usually stuck together on the most important issues that came up, we made no effort to agree on everything. We retained our differences in temperament and training—though we tried to subordinate these to the good of the team. And this was natural. We realized that we could get pretty tired of each other. We tried to avoid going around like a patrol of Boy Scouts or "Those Seven Little Dwarfs from Mercury." We tried to behave, instead, like seven vice presidents of a company. Each of us took charge of a different department. But we did not choose a president. We remained dependent on each other simply because we did have different specialties to take care of, and different functions to perform. Some of our functions overlapped. Take my own specialty, the pressure suit, for example. Obviously, you cannot consider the suit without taking into account the layout of the capsule in which the suit has to fit—which was John Glenn's area; or how you are going to get out of the capsule with this bulky suit on—which was Al Shepard's responsibility; or whether a man who has his arms and hands jammed into the sleeves and gloves of the suit can handle the controls—which were in Gus Grissom's department; or whether you can fit *two* microphones into the space helmet to handle the communications that Scott Carpenter had to worry about. All of these things had to fit together and be compatible. In a complex program like Mercury, everything crosses over and meets everything else on the other side. We all tried to avoid nit-picking with each other on these things. But we *never* avoided getting into technical arguments with each other when there was something big at stake. Sometimes, if one of us had an opinion that the others did not happen to share, we could really fight it out. We could not be expected to agree on everything, of course. And NASA wanted our different views. We also never worried about being polite. There was no time for that. We were big boys, and we could take the loud voices.

But through it all we remained good neighbors. We have dug postholes for each other's fences. Our wives are all very warm friends. And, like all pilots, we feel a strong sense of responsibility for each other. If one of us were to go down somewhere in a plane one dark night, you would find all six of us spending the entire night out there, trying to find a way to help him. We are still highly competitive, and we all want to beat each other to the good missions. Any pilot would. But we are also like the mountain climber who wants to be the first man to scale a peak where no one else has ever been. He knows that even a mountain climber cannot go it alone. He needs a team of other experienced climbers to back him up. So do we.

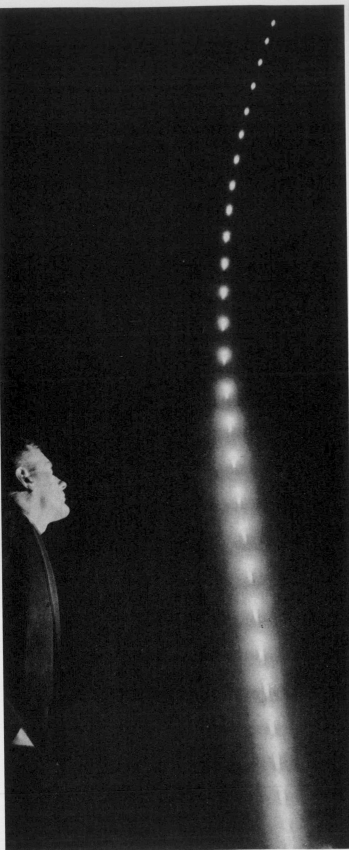

Deke Slayton specialized in learning about the Atlas missile which would put the Astronauts into orbit. In the multiple exposure at left he watches an Atlas being launched. Above, at the launching pad, he talks with B. G. McNabb, manager of the Atlas complex at Canaveral.

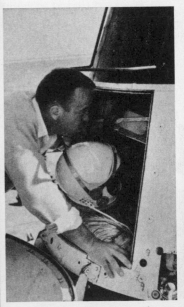

Alan Shepard helped work out recovery procedures to get the Astronauts safely out of their capsule after landing.

Gordon Cooper was responsible for checking production on the Redstone missiles that would launch Alan Shepard and Gus Grissom on their ballistic flights. In the tail section of a Redstone [LEFT], he looks over construction details with Dr. Joachim Kuettner. He also took on the job of perfecting a special knife for use as a survival aid in emergencies. In the workshop [CENTER] of Walter Randall, the knife expert who designed and manufactured these knives, Cooper tries out a prototype. John Glenn helped plan the layout of the Mercury cockpit and was especially concerned about the accessibility of switches and instruments. Here [RIGHT], dressed in his space suit, he tests his ability to reach all parts of the instrument panel.

Above, Shepard stands by and shouts last-minute advice through a porthole as Deke Slayton slides into a capsule in a rehearsal at sea. Frogmen bob around, ready to help the Astronauts practice their egress, and Wally Schirra crawls out of a toppled capsule into an emergency life raft as Navy personnel stand by.

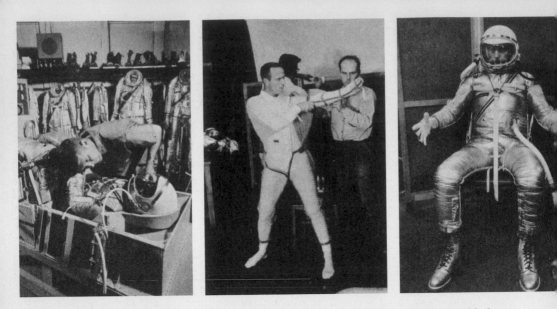

Wally Schirra's specialty was the environmental control system which provided an artificial environment inside the capsule and would keep the Astronauts at the proper temperature and under the correct pressure. A vital part of this system was the space suit itself. In the first picture above, Schirra bends over an uninhabited suit which has been inflated with air, as he listens for leaks. In the next two pictures, Scott Carpenter dons the special undergarment and then flexes his arms to test the complete suit for maneuverability. The hose leads to an air conditioner.

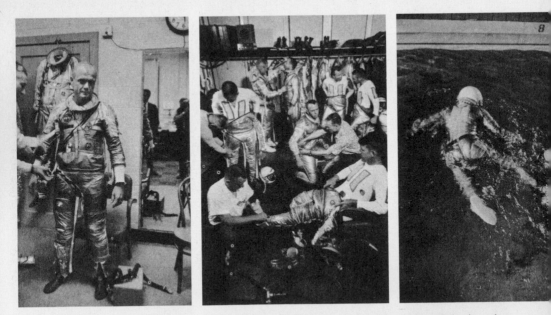

In the pictures above, John Glenn has his suit adjusted by Joe Schmitt, the NASA technician who dressed the Astronauts for their flights and helped plug them into the capsule. On the floor is a portable air conditioner which the Astronaut carried with him to keep himself cool inside the hot suit until he had entered the capsule. The entire Astronaut team suits up in the center picture, each man in his own tailor-made garment. Scott Carpenter, wearing his suit, splashes about in a pool while John Glenn, in diving gear, tries to pull him under to test the suit's buoyancy.

The Astronauts line up during a fitting session. In the front row are Schirra, Slayton, Glenn and Carpenter. Behind them are Shepard, Grissom and Cooper.

SEVEN MILES
AND A

John H. Glenn, Jr.

OF WIRE-
SWIZZLE STICK

I HAVE HEARD it said that no one but Rube Goldberg himself could have dreamed up the Mercury spacecraft that we have flown. If these people mean that it is a silly machine, they are wrong. If they mean that it is *complicated*, they are absolutely right. It is perhaps one of the most complex machines ever built. When you consider that this small spacecraft, which is not much bigger than a standard telephone booth, has thousands upon thousands of separate parts in it and seven miles of wire winding back and forth, linking all of these parts together, you get some idea of the intricate problem of designing and building it and making it work.

It is all the more amazing when you consider that the engineers started from scratch, and that they had a number of factors working against them from the start. For one thing, if we had had a more powerful booster we could have jumped a few steps at the beginning and could probably have done the job faster. That was the key to our problem—the thrust of our boosters. But we did not have a bigger booster at the time. We had a few coming along for the future—the huge eight-engine Saturn, with a combined thrust of a million and a half pounds, for example, and the new Titan II ICBM, which has a thrust of half a million pounds. But these were all in the future when Project Mercury started, and we could not wait for them. We were forced to use what was available at the time. We could not count on any breakthrough. So far as the booster was concerned, this limited us to the only big one we had that was about ready, the 65-foot Atlas ICBM, which has a total thrust in its engines of 360,000 pounds. This was just enough power to put into orbit a payload—including the spacecraft, the man and all the things he needed to control his flight and to keep himself alive—that could weigh no more than two tons.

The engineers had several limitations which they had to keep in mind from the very beginning. They had to move as fast as they could; they had to use "off-the-shelf" items in order to save time; they were also restricted in the amount of electrical power we would have on board to operate the systems, and the total weight of the capsule was severely limited. This last problem meant that our engineers had to take all of the gear that was readily available on the shelves—including pumps, radios, batteries, valves and nozzles—and miniaturize these parts wherever possible until the total weight of the entire capsule came within the limit. This was extremely difficult to do, but it was vital. For every part that we absolutely needed which was overweight we would have to leave another part behind that we decided we did not need quite so much.

It became even more complicated than that, for it required additional time just to miniaturize the equipment. The engineers needed eighteen weeks, for example, to devise a lightweight, spherical tank about the size and shape of a volleyball, which would contain the four pounds of oxygen that we would need on an orbital flight. The tank had to be strong enough to keep the oxygen under a confining pressure of 7,500 pounds per square inch, which was required to make the oxygen system work. We needed to install two of these tanks in every spacecraft—one for normal use, and one as an alternate to back up the first one in case it failed.

This—the backup principle—was one of the major headaches for the engineers, and one of the biggest time-consumers. In the interests of safety the entire system had to be studded with alternates or stand-by components which would take over in case the first or primary component failed. The same principle applied to all the major systems inside the spacecraft. The attitude control system, for example, was rigged in such a way that it could be operated either automatically by the capsule's automatic system, or manually by the gloved right hand of the Astronaut. These two alternate systems had to be kept completely separate from each other so that a fuel leak or a similar failure in one of them would not disable the other. The engineers worked out three different ways of firing the small retro-rockets which would start us back towards the earth when it was time to come home. These were wired so they could be activated either by the automatic timer on the instrument panel, by the Astronaut himself if the timer failed, or by command sent up by radio from control stations on the ground, if the other two methods failed.

The engineers had a word for this insistence on inserting backups into the system. They called it the principle of "redundancy." We felt that anything that was really needed to insure our safety on a flight was not exactly "redundant." But we agreed to go along so long as the engineers *meant* to say that their redundancies were imperatives. The important thing was that this doubling up of components added still more pounds to the crucial weight of the capsule and placed an even higher premium on miniaturization in order to make room for all of them. You can begin to see why there were a number of problems in putting it all together.

This was still not the whole story, however. On top of everything else, the engineers had to be sure that the entire system would perform automatically. This, again, was in the interest of safety. The plan called for sending up empty capsules before a human pilot would be risked—both on the Redstone training flights and on the Atlas orbital missions. This meant that the spacecraft had to be crammed with a fabulously intricate set of timers, sensors, self-regulating controls and automated valves which were capable, if necessary, of flying the spacecraft all by themselves. They would regulate the flow of oxygen through the cabin and the Astronaut's suit, maintain the cockpit at the prescribed pressure and temperature—*and* report down to earth on how all of these procedures were being carried out so that if the flight failed the engineers would know exactly what went wrong and could remedy the failure before taking unnecessary chances on the next flight. It took added months of effort to devise these measures and get them properly installed in the system. They also added more pounds to the total weight and took up space that was already critically limited.

Then there was the largest problem of all, so far as the capsule was concerned. In addition to testing all of these components to make sure they would operate under the stresses of vibration and heat that we would encounter, the engineers had to see to it, somehow, that all of these efforts came out even with respect to time. At the same time that the spacecraft was being assembled, NASA was also checking on a number of other important items that were coming along in different places—the boosters and rocket engines, for example, which were being manufactured and tested especially for Mercury, the extensive communications and tracking facilities which were being set up at isolated stations scattered around the world, *and* the Astronauts themselves. We were the human cogs that had to be inserted into the system at the proper time. So we had to follow a schedule in order to get all this done on time and see that it dovetailed. The schedule was flexible. We knew that variable factors such as weather, over which we would have no control, could cause delays. Someone, however, had to at least guess at a good schedule well in advance and try to get everyone to meet it, for there were other participants in the program outside of NASA's direct control who needed time to get ready. The U.S. Navy, for example, with its fleet of recovery ships, would have to be alerted about launching dates far enough in advance in order to provide the necessary support and still perform its other important duties.

All of this called for a tremendous display of coordination and close timing. The U.S. missile program was an earlier example of a similar problem. With the Soviet Union already building ICBM weapons and presumably installing them on launching pads, complete with trained crews, the U.S. had no time to lose trying to catch up. So while early American missiles like the Thor and Jupiter IRBMs were still being tested for reliability at Cape Canaveral—and occasionally misbehaving—the U.S. was already hard at work training the men who would be assigned to launch them in the event of war; at the same time it was designing the launching pads to hold them so that everything would be prepared ahead of time for the arrival of the boosters themselves.

The scheduling problem which the U.S. undertook with Project Mercury was of a far different magnitude, of course. There were so many unknown problems involved in Mercury and so many "firsts" to be accomplished, that no one could be certain, when the project first began, just how it *would* come out. No one knew with what precision, for example, the spacecraft and the Atlas would fit together or "mate" when the time came. Certainly no one knew at the beginning just how well a man would fare on the kind of ride that this combination of spacecraft and Atlas would give him. This was the purpose of the exercise—to find out.

One of the first things that had to be decided was what kind of spacecraft to build. The size of it was already more or less dictated by the load that an Atlas could place into orbit. But what shape should it have? Spherical, like a satellite? Cylindrical, like a rocket? With wings, like a glider, to bring it back?

All of these ideas were discarded almost from the start. The shape that finally won was the brainchild of a NASA engineer named Maxime

Faget. Max had been working on shapes for missile nose cones that had to hold together as they descended through the friction of the atmosphere and he adapted some of these ideas for the basic design of the spacecraft. He came up with a rather squat object with corrugated sides, a smooth rounded bottom, and a round top that looked a little like the cap on a catsup bottle. The whole thing resembled a gigantic TV tube. It turned out that the McDonnell Aircraft Corporation in St. Louis had been working roughly along the same lines. So, when the time came to get serious and start building hardware, McDonnell was ready to put together some models which Faget and his engineers could test. They put the shape through more than one hundred strenuous tests in some thirty different wind tunnels across the country. These tests indicated that the shape would withstand the tremendous aerodynamic stresses of an Atlas launch into space. They also proved that the spacecraft could perform a very important function just because of its peculiar configuration. Its unstreamlined shape would provide enough "drag" to slow it down on the ride home. This proper balance between speed and friction would keep it from burning up. Just to make sure, NASA engineers burned models of the capsule in a 6,000-degree jet flame. They dropped it into water and sand, and onto solid concrete to find out if it could withstand the shock of an emergency landing on the earth. Since the body tissue in pigs is very similar to that of humans, they also strapped a live pig in a contour couch, loaded it aboard a capsule and dropped the capsule down a shaft to demonstrate that the shock of a hard landing would not necessarily kill a living passenger.

The shape was finally approved. Then McDonnell had to figure out a way to build it so it would be as strong as possible and as light as possible at the same time. The engineers knew that every pound saved on the pad would provide an additional mile in range.

The wall of the capsule is made up of two layers of high-grade metal. The outer layer consists of shingles made from a metal called René 41 which have been corrugated and then welded together to give them extra strength. The welding technique had to be specially perfected so that the thin sheets of metal would not be torn or cracked in the process. The inner layer is made of titanium, a light, strong metal which was developed for jet engines and provides the strength of steel at about half the weight. The two layers are separated by a hollow space that provides extra insulation.

It was an extremely difficult vehicle to build, and it was full of compromises. It was not perfect, but it was functional. It would fit on top of a Redstone or an Atlas, and it was just large enough to contain a man and all the accessories he would need to function as he started his investigations of space. We sometimes joked, "You don't climb into the Mercury spacecraft, you put it on." You squeeze past all the gear that is mounted inside, like a man sliding under a bed. Once inside, you almost feel like just one more piece of equipment—the most important piece, of course.

As you slide into the cockpit, the first thing you see is the instrument panel. It is right in front of you, and every available inch of it is crammed with meters, switches, toggles, buttons, levers, dials and lights which you will have to monitor or operate from the moment you settle down and buckle your straps until you are ready to explode the hatch and leave the capsule at the end of the mission.

This is the cockpit, the real "opinion area" of the capsule, which had been largely designed before we got into the program. Our views were respected, however, and some changes were made in the basic layout to accommodate our ideas. The cockpit looks fairly formidable, even to an experienced jet pilot—especially if he happens to be a creature of habit and is used to finding his fuel gauges and battery switches somewhere else. That was one reason why we had some trouble getting a total agreement on how it *should* look. We all had varying ideas. Once we got used to it, however, the layout made a lot of sense. Since we were using many off-the-shelf items, it did not look unlike a standard cockpit. We tried to make everything easy to see at a glance. When you are being pressed back hard into your couch by the acceleration of the launch, you are not able to turn your head very far to see the instruments. You are fairly rigid, and you have to scan the panel by moving your eyes, not your head. The panel is 24 inches in front of your face, which is close. We asked the engineers to use different colors to help us spot various instruments at a glance and we grouped the buttons and lights and switches together as much as possible according to their special functions. The panel on the far left, for example, holds most of the miscellaneous controls which the pilot has to reach for in flight while he keeps his other hand, his right hand, busy with the attitude-control stick. There are handles for turning on the manual yaw, pitch and roll controls—you push them to switch the system on, pull them to switch it off. You superintend the retro-firing sequence from here with toggle switches. You have one large handle for repressurizing the cabin with oxygen in case of a bad leak, plus another handle to decompress it and get rid of the oxygen in case of fire. Decompression—or getting rid of the oxygen—is the quickest way to put out a fire, incidentally, since the vacuum of space would not provide enough oxygen to support combustion. On the next panel to the right is a series of lights, all in the proper sequence, which indicate to the pilot whether various functions which are supposed to take place automatically have actually occurred—such as jettisoning the escape tower when it is no longer needed, or separating the capsule from the booster at the moment of insertion into orbit, or breaking out the parachutes to bring you to earth. The light is green if the function has been performed. It is red if there has been a failure of some kind. At the left of each light is an "override" switch which allows you to take over from the automatic system and do the job yourself. A bright red light at the very top of this panel is labeled "Abort." It would come on during the launch phase as a warning if something had suddenly gone so wrong with the capsule that you would want to get out immediately. You could be aborted automatically at this point, or by a command from the ground, or the Astronaut himself could start the sequence that would fire the escape tower to pull the capsule free from the booster. There is also an "override" switch connected with the abort light, so you know that if you should be aborting automatically but nothing is happening, you can take over and perform this vital function yourself.

In the center of the panel is an array of dials to give you information about your flight. You can tell by glancing here how fast you are going up—or coming down—what your altitude is, and how much hydrogen-peroxide fuel you have left to run your attitude-control system. At the very top of this section is a series of needles that tell you exactly how far—if at all—you are yawing, pitching or rolling. We worked out this display after several trial-and-error experiments to find the right combination which would give us the most information in the clearest manner. Between your knees at the very bottom of this display is the large periscope which gives you a horizon-to-horizon view of the earth below so you can check your actual attitude against the needles. The eyepiece of the periscope is eight inches in diameter. A magnified section in the center of it, which gives you your most exact picture, is five inches across. You have a 190-degree field of vision. The window we requested is located above the display of attitude needles and gives you an even better view of the reference points you need. There are two other instruments on this panel. One is a dead-reckoning device which we call the EPI (for "earth path

indicator"). It shows you—on a small scale model of the earth that re-
volves as the spacecraft moves along in orbit—exactly what section of the
earth you are passing over. The other is a "satellite clock," which is really
four clocks in one and performs four separate functions. One clock shows
you what time it is according to Greenwich Mean Time, which is being
used to coordinate the mission in the Control Center at Cape Canaveral
and at all the tracking stations on the world-wide net. A second clock,
which you use to time all the events that will take place on the flight,
shows you how many hours, minutes and seconds have elapsed since lift-
off. Another clock runs backwards and shows you how much time you
have left before the firing of the three retro-rockets that will start you
back to earth. The fourth clock actually fires the retros when the time
has come. This clock is preset on the ground according to a timing for
retro-fire which we have computed before the mission. It can be reset
during flight, however, either by the Astronaut or by a command from
the ground if it turns out that we have gained or lost a few seconds during
the flight and would come down at the wrong spot if we did not alter the
timing. We also have a stop watch in the capsule; and all of us wear very
exact watches on our wrists. As you can see, we are extremely time-con-
scious during a mission. We have to be. For we are flying at a speed of
about five miles per second, and if the retros fire one second too soon or
too late we can miss our landing area by as much as five miles. If that happened, the ships might have a little trouble getting to us.

The next panel to the right—up at the very top—shows us all we need
to know about the environmental control system which keeps us alive—
what the temperatures and pressures are in the cabin and in our suit.
Down below this is an array of meters which monitor the electrical equipment in the capsule. At the bottom of this section is a panel full of controls
for the communications system that keeps us in constant touch with the
ground—either by voice radio or by coded dots and dashes. At the extreme right is a panel which we did not have in the capsule when we
started—and which Al Shepard did not have in his cockpit. This panel
groups all of the warning lights in one convenient place so we can see at a
glance if any problems have cropped up. Each light indicates that some
particular failure is about to take place in one of the systems, such as a
dangerously low level of oxygen remaining which might cause us to make
an early re-entry. At the extreme edge of this panel is a series of fuse
boxes. These were located in several places when the capsule was first
designed, but we asked for them to be brought together so we could see

them at a glance and make a fast change of fuses if one of them happened to blow. Being able to change fuses in mid-flight and keep the electrical circuits clear gave us an extra backup that we did not have when we started the program.

The pilot has to monitor all of these gauges and lights continuously. He must be ready to move instantly to any switch on the panel if he sees a red light go on, or if he notices that a green light has failed to light up when it should have. This is one big reason a man is up there in the first place—to make *sure* that this machine works and to make the necessary corrections if something goes wrong, which could save us millions of dollars and months of effort. All of the controls are fairly accessible, and during a normal flight, even when we are pulling Gs, we do not have much trouble reaching for them and making contact. There is one condition of space flight, however, in which we might have trouble. That would come if we suddenly had to pressurize our suit because we had lost cabin pressure. A fully pressurized suit is a clumsy affair, and you can have trouble moving your arms about inside it as easily as when the suit is not pressurized. To prepare for this eventuality, we have added a simple little rod to the other equipment on board. It is ten inches long and has a hook on the end of it for pulling at levers and a stub for pushing at buttons. You grasp it in your glove if you know you are not going to be able to reach something with your fingers. It is really an extension of the finger. We call it, naturally, a "swizzle stick."

OUR COZY

Walter M. Schirra, Jr.

COCOON

THERE WAS NOT a single item in the capsule, from the swizzle stick to the fuse boxes, that we did not need. But if we ever had to make up a list of the items that we could throw out last, I think most of them would probably fall into my area—the life-support or environmental control system.

The first thing you do after you have crawled into the capsule—or "put it on," as John Glenn puts it—is to slide, bottom first, into the custom-made couch. The couch has been made especially for you and squeezed into the cockpit like a shoe-brace into a shoe. You strap yourself into it with a special harness that keeps your shoulders, chest, waist and legs pinned in place like a doomed man in an electric chair. Each of us has his own preferences when it comes to the straps. We all have different ideas on how they feel when they cut into our bodies, or how tight they ought to be. And we do not all wear the same combination. There is nothing optional, however, about the couch. Outside of the fit, it is the same for all of us. It is a kind of sandwich made of crushable, honeycombed aluminum which has been bonded to a fiber-glass shell and lined with a thin layer of foam-rubber padding in an attempt to make it comfortable. We had to sacrifice real comfort, however. For if the padding were too thick, we would tend to bounce up and down on it and could set up a rebound effect that would be even more uncomfortable. When you get out of the

capsule after spending several hours in it, you have stiff muscles and feel
a slight dizziness from lying with your legs elevated. This position results
in the blood pooling in the upper part of your body and it is one reason
why we look forward to going weightless. This relieves the cramped
feeling and the pooling of the blood.

Each of us had his own couch molded to the back of his body—with his
pressure suit on—so that the protective recess we crawl into exactly fits
the sloping lines of our shoulders, the contour of our backside and the
length of our legs. Sliding into the couch is a little like slipping a ginger-
bread man back into the cooky cutter. Since we are all different-shaped
cookies, we need our own private little recesses to lie in.

Originally, the engineers thought that a couch made of a weblike net-
ting, like the material in a hammock, would offer the best protection. It
would give way slightly as we were forced back into it during the periods
of high acceleration, and it would help ease the stress. This theory
turned out to be wrong, however. We discovered that what we really
needed under high G stresses was very solid support to help our bodies
fight back against the G forces and operate efficiently. When a man is
pulling heavy Gs in the normal sitting position, the blood drains down
from his head, and he can have a gray-out or a black-out—or even die if
the force is strong enough. The blood that drains down tends to pool in
the abdomen. But by designing a couch for us to *lie* in, with every portion
of our bodies resting on a solid surface and with our legs raised up on a
level with our heads—the way you lie back in a contour chair—we were
able to minimize this danger. The couch we got was a fine compromise.
We used these same personalized couches when we went for training
rides on the centrifuge, and we kept them stored—with our names
stamped on them—so we would be sure to get the right one whenever we
were getting ready to undergo much stress.

The couch was another one of Max Faget's contributions to our welfare, and there is an interesting story connected with it. Some military
doctors who had been studying the effects of G forces on pilots had
warned NASA to come up with a capsule which would subject us to no
more than 12 Gs, either on take-off or on re-entry, since this was assumed
at the time to be the maximum G force that a man could normally survive.
Faget and his assistants worked it out mathematically that the capsule
they had in mind would subject a man to no more than 8 or 9 Gs if everything worked properly. So it all seemed fine at first glance. But then the
engineers computed that if an emergency cropped up just before the

capsule went into orbit and it had to be tugged away from the Atlas by the escape tower in a sudden abort maneuver, the strain of this sudden increase in motion might subject us to 20 Gs or so for almost a second. Worse than that, the medical advisors predicted that if the capsule happened to re-enter the atmosphere at a faulty angle and started plunging down too steeply instead of following a gentle curve, we might encounter a force of 20 Gs that would last up to eight seconds.

The doctors decided that a man would probably survive this experience. But he would be pretty well churned up, and it might take a while before he could get out of the hospital. As Faget said, "You can't plan a flight program with possibilities like that."

So the engineers started looking for a solution. For a long time, they could think of nothing. They had just about decided they would have to go back to the drawing board and come up with an entirely new concept for the capsule when Faget suddenly had a hunch and told two of his engineers to put together a solid, form-fitting couch. He was not sure, but he thought it would work. Then he asked a NASA test pilot named Robert Champine to try it out on the centrifuge at Johnsville. This was the huge wheel-like device with an arm jutting out like a spoke that could whirl a man around at speeds up to 48 revolutions per minute and subject him to forces as high as 40 Gs. We Astronauts later become regular passengers on this centrifuge when we got deeper into our training. But the experiment Faget had in mind was a stab in the dark. Champine climbed into the cab on the end of the centrifuge arm and lay pressed into the couch while the wheel started up. He was spun around and around, until finally he reached the force of 12 Gs before the machine was slowed to a stop. Champine said he felt fine and was willing to try for more. But he was scheduled to attend an urgent conference that had already been set up on the West Coast, and he had to leave. At this point a Navy lieutenant named Carter Collins, who was two inches shorter and 30 pounds lighter than Champine, volunteered to ride in his place. The engineers had to pad the molded recess in Champine's couch with extra foam rubber to accommodate Collins' smaller frame. But Collins got the same ride that Champine had taken—up to 12 Gs—with the same good results.

"Let's see how high we can go," Collins suggested when his first ride was over. Max Faget, who had come over to Johnsville to watch the demonstration, took him up on the offer.

"If you can get up to 20 Gs," Max said, "you will be my hero for life."

The next day Collins slid back into the couch. This time the centrifuge spun faster and faster each time it went around. First, Collins rode to 13 Gs, then to 14, then 15, 16, and 18. On the sixth and final run, Faget got his hero. Collins pulled a total of 20.4 Gs for more than six seconds, and he came out in good shape. The Navy doctors who examined him did discover that some tiny blood vessels under the surface of his skin had been ruptured as a result of the tremendous pressures, but this was not a serious injury and it did Collins no real harm. Collins proved that the couch worked; the problem of overcoming re-entry G-forces was solved. Later on, when a live pig was dropped inside a capsule—because a pig's weight distribution is similar to man's—the pig was strapped to a contour couch, too. Its safe fall to earth helped confirm the principle; the couch would also take up the shock.

The finished couch was the product of some careful and clever engineering. But it was a simple piece of furniture compared to the lash-up of tubing, fans, filters and tanks which was built around it and which we called the ECS (environmental control system). The ECS stretches through every nook and cranny of the capsule, and its function is to provide us with a man-made environment which will keep us alive and comfortable and help us to work efficiently in the vacuum of space. The entire ECS layout weighs only 85 pounds—it is a masterpiece of miniaturization—but it can perform an amazing number of vital chores.

The centerpiece of the entire system is our pressure suit. This is the skintight cocoon of rubber and fabric into which we seal ourselves before each flight or practice mission. Essentially, the suit is a man-shaped balloon. It is also tailor-made, and it fits us so snugly that we have to use thirteen different zippers and three rings to put it on. Complete with helmet, the suit weighs about 22 pounds, and each one costs about five thousand dollars. Each of us has three space suits—one for "everyday" wear on our training missions, a second, or "Sunday-go-to-meeting" suit, which we reserve for our trips into space, and an extra Sunday suit which we keep on hand on the launching pad as a backup for our flights.

We wear the suit throughout the mission. But because the capsule itself is pressurized, we do not necessarily have to keep it sealed and inflated all of the time. We can open up the visor on the helmet, for example, when we want to grab a bite to eat or scratch our nose. But if the cabin itself springs a leak and the pressure begins to fall, a system of delicate barometric sensors spots this change in pressure immediately and signals us to close up the faceplate immediately. Closing the visor reseals the suit and it inflates automatically so that we are under a safe pressure no matter what happens inside the cabin. The suit, then, is a kind of cabin-within-a-cabin or a portable home-away-from-home.

I acted as a kind of consultant tailor on the pressure suit. It had a number of things wrong with it when we first started the program. But fortunately we had time to play with it—in fact, we nit-picked it to pieces—and we wound up with a very useful gadget.

Because space suits are personalized garments, you need to make more alterations on one of them to make it fit properly than you do on a bridal gown. It is not possible just to walk in and buy a pressure suit off the rack. In order to get the precise measurements for ours, we visited the B. F. Goodrich plant in Akron, Ohio, where the suits were made, and stripped down to our long johns so the technicians could plaster us all over with strips of wet paper. The paper then dried into a mold which Goodrich used to form the suits. It took them about a month to complete each one. Then we went on tinkering with the suits and adding more refinements until the ones we have now bear only a general resemblance to the models we started out with.

For one thing, we had to make them easy to wear. Even when it is not inflated, a pressure suit is a cumbersome garment. It has an inner layer of rubber and an outer covering of nylon which has been aluminized to reflect the heat and radiation and help us keep cool. When this double layer is blown up to its full pressure of five pounds per square inch, the suit is downright rigid. As Al Shepard's recorded friend, José Jiménez, says: "It is very uncomfortable."

Since we had to breathe and work inside this cocoon, one of our major problems was to make it as pliable and comfortable as possible. If the suit got so tight that it resulted in cramps and pressure points, we might even have to abort the mission. The people from NASA, Goodrich and the Navy solved this problem in a number of ways. The original plans, for example, called for an extra layer of sponge-rubber insulation in the suit to help protect us from the heat during re-entry. After we discovered from some unmanned flights, however, that the cabin would not get nearly as hot as we had thought it would, we decided to eliminate this extra insulation and cut down on some of the bulk.

We also had several problems with our gloves. The first suit that we tried called for straight-fingered gloves which were locked onto the sleeves with zippers to make sure that the pressure did not leak out through our arms. The trouble with this arrangement was that when we tried to clench our fingers inside these pressurized gloves—in order to grasp the control stick, for example—we had to exert a continuous effort which could get very tiring. Then someone made a startlingly simple suggestion: Why not form the gloves with fingers that curved inward in the first place? This way we would not have to exert any effort at all to cup our hands around the stick. We tried the idea out. But this presented us with another problem. Some of the buttons on the instrument panel are set into little recesses so that you cannot push them or brush against them inadvertently and foul up the sequences. With out fingers curved, we were unable to poke into these recesses. We solved this dilemma by weaving one finger on the left hand—the middle one—so that it stood out nice and straight and provided us with a fine tool for pushing buttons when we did not want to bother reaching for the swizzle stick. Another problem involving the gloves was that they sometimes slipped off the controls. This was because we keep the air fairly dry inside the capsule, and this tended to make the leather gloves slippery, like skates on ice. We solved this by weaving some extra threads into the material which made the texture rougher than you normally find in gloves—a bit like sandpaper, in fact. This provides a sort of nonskid tread for our hands and makes it much easier for us to push buttons and flick switches. One final problem with the gloves was the zipper which bound each one of them to the sleeve. The zipper arrangement made such a stiff, unbending unit of our wrist and arm that we had trouble twisting our hands to manipulate the controls. We took care of this by inserting between the gloves and the sleeves a rotating metal ring that swivels freely and lets the hand turn independently.

This took care of our hands. Another part of an Astronaut's body which juts out of his suit and has to be sealed inside it is his head. The space helmet takes care of this. Like the suit, it has to be a perfect, individual fit. We all flew to Los Angeles, where the liners for the helmets are made, and had a cast made of our heads to form a mold for them. The liner has an outer covering of leather and two layers of foam inside it—a thin layer of soft "comfort foam" and a thicker layer of stiff "crash foam" to hold the helmet firmly in place on our head. The liner itself fits inside the outer shell of the helmet, which is made of hard fiber glass. The crash foam is designed to give way very slowly under impact. This protects and cushions the head better than something that is too soft, for the soft material would give way too quickly under a hard knock and might start our head bouncing and reverberating around inside the hel-

met. We have discovered, incidentally, that the entire rig fits best when
we have about a five-day-old haircut. John Glenn refined this a bit and
had his hair trimmed every three days before his flight.

We tested the suit and helmet in every way that we could think of.
We baked them in heat chambers where the air got up to 180 degrees
and the walls got up to 300 degrees—which is warm enough for broiling.
We wore them on the centrifuge to find out how well we could operate
in them under great acceleration. And some of the factory models were
pressurized to the bursting point—which does not occur, incidentally,
until the suit is under four times as much pressure as we will get on any
of our missions. We just wanted to make sure. We also made a few fixes
to accommodate one or two of the Astronauts who had a special complaint. We changed one of the zippers to one side, for example, and
split it up into two separate zippers because it pressed right against
Deke Slayton's Adam's apple and we were afraid it might be rather pain-
ful for him on a long ride.

Normally, unless the cabin pressure fails and we are threatened with
a case of the bends—which would almost certainly be fatal—the suit serves
mainly to keep us ventilated and provides us with a constantly refreshed
supply of oxygen to breathe. It is a beautifully contrived mechanical
environment. A steady stream of pure oxygen—which emanates from one
of the two volleyball-size flasks we have mentioned earlier—enters the
suit through an inlet valve near the waist. We plug this valve in as
soon as we enter the capsule. The oxygen then circulates through the
entire suit and reaches all of our extremities to cool them off. A series
of waffle-weave patches on our long-john underwear helps to keep the
oxygen moving. The oxygen finally makes its exit through an outlet located near our right ear in the helmet. As it flows out, the oxygen takes
with it whatever body odors, perspiration, carbon dioxide, water and
other waste matter—like nasal discharge or bits of hair—it has picked up
on its tour. The water and CO_2, of course, are normal by-products of the
pilot's metabolism. The system dumps all of this into a marvelous bit of
plumbing which traps out the waste and uses an electric fan to push the
tired gas over a bed of activated charcoal. This filters out the odors and
then the gas goes on through two tanks of lithium hydroxide to remove
the deadly carbon dioxide. When it has been thoroughly cleansed, the
oxygen then goes into a cooling device which removes the excess heat
which it has picked up from our bodies. The hot gas is cooled by con-
duction in a heat exchanger. Here water takes up much of the heat, which

is then discharged out of the capsule in the form of a puff of steam. One of the many things we have to worry about on a flight, incidentally, is making sure that this steam duct does not freeze up because of excess water flowing through the system. If that happened, the water would not get converted readily into steam, the steam duct would clog up and cease functioning, and so would the delicate systems which keep the Astronaut and his cabin cool. A device which Rube Goldberg himself would be proud of takes care of droplets of water which collect in the suit circuit at this point. The drops are soaked up by a little sponge which is struck by a small piston every thirty minutes to squeeze the water out and dump it into a storage tank. This, incidentally, is our emergency supply of drinking water. As soon as the oxygen has been purged of water, waste, odors and carbon dioxide, it flows back into the suit and starts the same tour all over again. While all of this is going on, a much simpler process is taking place in the separate atmosphere of the cabin. Here, a fan and a heat exchanger keep the cabin ventilated and cool. If the oxygen starts leaking from the cabin system, it can be replenished from the oxygen in the suit—after the Astronaut has had a chance to breath it first.

Oxygen for both the suit and the cabin comes from the same source, but the flow is controlled in such a way that if the cabin springs a serious leak and the pressure falls, the flow of oxygen into the cabin is cut off and the supply is saved for the suit. If it is the suit that fails us, we can open up the faceplate in our helmet and breathe the cabin air. If they both fail at the same time, we have an emergency supply of oxygen which will keep us alive long enough to wind up that particular orbit and make an emergency re-entry. If all three systems go out, including the backup supply, we are dead. We are fairly certain, of course, that such a disastrous thing as this will never happen. The odds are definitely against it.

We all like different temperatures in our little homemade environment. The system was designed to keep us at anywhere from 50 to 90 degrees Fahrenheit, and we have the option of picking the temperature that seems the most comfortable and efficient for each one of us. Al Shepard likes his oxygen cold—about 60 degrees. I like it to be at about 68 degrees when it comes through the inlet into the suit. Some of the other fellows go as high as 72 degrees. We do not have any option, of course, on the pressure that we keep inside the cabin. Before launching, the capsule cabin is purged of ground air and filled up with pure oxygen. Then, after the launch, the cabin is automatically sealed off and held at a pressure which is five pounds per square inch greater than the pressure outside, no matter at what altitude. Out in pure space, where there is no outside atmospheric pressure, this means a cabin pressure of exactly 5 psi. At 27,000 feet, where the atmospheric pressure is 5.5, the cabin pressure works out to 10.5 psi. In effect, this makes an inflated balloon out of the capsule—with more pressure on the inside than out—and helps to strengthen it. The atmospheric pressure at 27,000 feet is only one-third of what it is at sea level. It is just about what you would find along the rim of the Himalayas. But it is livable. This is because the atmosphere inside the cabin is made up of 100 percent oxygen. At 27,000 feet, the concentration of oxygen in the atmosphere outside is less than 7 percent.

Once we are up, and if both suit and cabin systems are working, we can open up our visor and breathe the cabin air for a bit and eat a bite of food. We would not be able to eat, of course, if the cabin pressure went out and we had to stay sealed up inside the suit and helmet. We are supposed to keep the visor closed for re-entry, for this phase of the flight is the greatest test of the environmental control system. The walls on the outside of the spacecraft can heat up to as high as 3,000 degrees Fahrenheit during this period, and the inside walls around us could get as hot as 200 degrees Fahrenheit. If the cooling system holds up, however, we should feel no hotter on the inside of the suit than we would out-doors on a muggy summer day. In the final stages of descent, a snorkel opens automatically at about 20,000 feet, and brings in fresh, cool air from the outside which purges the hot suit and gives us our first whiff of the briny ocean as we settle down. We also switch on the emergency oxygen at this point to bring in a fresh supply.

It is quite a system, really. There is enough oxygen stored in those little tanks to keep us going for about 28 hours before it is used up. We were not prepared to stay in orbit that long on our first flights. But all that oxygen would come in very handy if we happened to get hit by a meteorite or spring a leak and our environment started to dribble away on us. If this ever does happen, the pilot can declare an emergency and draw on the backup tank with its additional four pounds of oxygen. This supply would also leak out, but it would be used up very slowly—at the rate of about five-tenths of a pound per minute. This would give us about 80 minutes of breathing time, enough to last until the end of that particular orbit. The snorkel would take care of us as soon as we penetrated far enough into the atmosphere to be able to breathe on our own.

We gave the ECS over 500 hours of tests on the ground—with a man breathing from it—before we took a chance on it with Al Shepard.

SOME
TRICKS OF

Malcolm Scott Carpenter

FABULOUS
THE TRADE

WITH ALL THOSE miles of wire and dozens of little switches tucked away in its innards, the capsule has a lot of interesting things to do besides keeping the Astronaut alive and comfortable. For one thing, it must also keep him in constant touch with the control stations on the ground. The Mercury staff back at the Control Center on Cape Canaveral must know what is going on, minute by minute, so they can make decisions that affect the mission. The aeromedical people, who are scattered around the world during an orbital mission to keep an eye on the Astronaut's physical well-being and detect any symptoms of trouble, have to have all of the facts in front of them before they can make a useful diagnosis. The capsule must be able to send this information down on its own if the Astronaut is unable to. Fortunately, our capsule is a talking wonder.

We can communicate from the capsule to the ground by both voice and Morse code, and we have a number of transmitters to do this with in case one of them fails. Like everything else in the capsule, the communications system has backups on top of backups. We have even gone so far as to develop a foolproof system which makes sure that no stray signal from the ground can sneak into our circuit and make the capsule do something that *we* don't want it to do. The hookup is designed, for example, so that the right people on the ground can trigger the retro-rockets to bring the capsule home. Or they can institute an abort if things suddenly turn to worms and the automatic system fails to take over. But the capsule simply won't accept wrong numbers or messages from the wrong people. It is just too smart.

In addition to the radios, which we can talk over or use to send coded messages, we have two separate telemetry links with the earth which broadcast a constant flow of facts about how the capsule and the Astronaut himself are doing. All of these little facts—like how hot the suit is, or how much oxygen is left in the primary system, or how much fuel remains in the manual control system, or how fast the Astronaut's heart is beating—all of these small bits of information are gathered up at regular intervals by an electronic device and then translated into blips of radio energy which are sent down to earth. When the blips reach the earth, they are translated back into the same facts again, which come out this time as moving lines of ink on long rolls of telemetry paper which the flight monitors can read for themselves.

Each time the capsule performs an important function—like jettisoning the escape tower, for example, or firing the retros—it automatically sends a telemetry report to this effect down to earth. Even if the Astronaut happens to be out of action or too busy with something else at that particular moment to report each fact, the men who are responsible for him know exactly how things are going and can jog the system with a radio command if an important event or sequence appears to be hanging fire.

The electrical energy to do all of this is provided by an array of batteries: three main batteries which are hooked up at the start with all of the circuits; two standby batteries which are separated from the main batteries so they will not short each other out; and an isolated battery which serves as still another backup. The electrical system is tremendously crucial. It not only backs itself up, but it helps us back up some other functions as well. If the capsule fails to separate automatically from the booster, for example—that is, if the explosive bolts which help hold the two components together fail to explode at just the right moment—we can release the capsule manually by flicking a switch. This sends out an electrical impulse which explodes the bolts and fires the posigrade rockets that shove the capsule and booster apart.

The three main batteries also power the air conditioner which keeps us cool, and the flashing lights on the instrument panel and the tone alarm in our headset which warn us if something is going wrong. They also provide the juice to work the hydrogen-peroxide nozzles which are mounted on the outer skin of the capsule to keep us in the proper attitude. In addition to all this, we also need power for the four cabin lights which help us to see our instruments when we are on the dark side of the earth —which is about half the time.

The brain of the capsule—which backs up the brain of the Astronaut and is backed up in turn by his—is the Attitude Stabilization and Control System (ASCS) which Gus Grissom helped to perfect. This device is crammed with sensitive gyros and horizon scanners which keep the capsule lined up at the proper angle with respect to the earth, and it manipulates a total of 12 attitude-control nozzles which are part of the automatic system. Some of these nozzles give us 24 pounds of thrust to make major corrections or changes in the capsule's attitude, and some put out only six pounds of thrust to make the more minor corrections. We can choose the thruster we want to use if we take over manual control. The ASCS is capable of some uncanny feats. For example, using the horizon scanners, which are infrared eyes that determine the position of the horizon outside, the ASCS senses automatically how far the capsule may have moved from its programed attitude in yaw, pitch or roll, then sends electronic impulses out to the solenoids to activate whatever fuel valves are needed to get the capsule back into proper alignment. The valves feed the nozzles which squirt out the decomposed fuel to make the actual corrections. I had a good deal of trouble with this device on my flight. It was supposed to keep me aligned automatically so that I would have more time to concentrate on other things, like observing the sunsets, the haze layer and the balloon that I was towing along behind me to test my ability to judge distance and distinguish colors in space. If the scanners are working properly, you should be able to look out the window and see the horizon about two-thirds of the way up the window. Just before it was time for retro-fire, I set the capsule up in the proper attitude three different times, and then tried to get some other things done. But each time I looked out the window, the capsule had pitched down again and I had to take over and line up the capsule all over again. This was disconcerting, and it took a lot of my time. We will not be using this same system on most of our future flights, however, so the fact that it did not work properly on my flight did not mean that we were in real trouble for the future and would have to go back to the drawing board.

The odd-looking tower on top of the capsule is also capable of performing some interesting stunts. This is the escape tower, and it could save our lives. It is 17 feet high, made of tubular steel, and is attached to the capsule by explosive bolts which blow apart and separate the two components when the tower has done its job and is no longer needed. Hopefully, the tower will never have to do its job. It carries a relatively small rocket which is loaded with only 285 pounds of solid fuel, and the rocket is designed to burn for only a little over a second. But in that brief moment the tower can pull the capsule away from the booster and carry it to safety if something has gone wrong. The rocket on the tower is angled slightly so that it pulls the spacecraft not only upward but off to one side so that it will be out of the way of the booster if it blows up or keeps on climbing. As soon as the tower reaches the proper altitude, it breaks out the parachute which lowers it and the capsule to earth. All of this takes place automatically and is geared to happen only in the case of an emergency. If everything goes well and the tower is not needed, it is jettisoned automatically at the proper time and falls back into the sea. The capsule is now on its own and about 600 pounds lighter for the long ride ahead.

This is not the only striptease which the capsule performs in space. At launch the capsule weighs 4,265 pounds. By the time it gets into orbit it is stripped down to 2,987 pounds. And when it is picked up at sea, minus its parachutes and recovery aides, it weighs 2,422. Another item which is shucked off, just before re-entry, is the package of three retro-rockets which ride strapped onto the heatshield during most of the mission. These rockets are also fairly small, and they consume only 144 pounds of fuel in the few seconds that they fire. But, by firing forward, in the same direction in which the capsule is moving, the retros kick up enough thrust to put on the brakes and slow the capsule down by about 500 feet per second. This is just enough to allow the force of gravity to get a good grip on the capsule and start pulling it back down into the atmosphere. The capsule should not re-enter too quickly or the deceleration will be too great. It should not creep back too slowly or at too slight an angle, either, for this would result in a long re-entry procedure which would build up too much heat. The angle must be precise—which means that the pilot must stay right on top of the attitude control system and have the capsule pointed exactly right when the retros fire. Here again, the capsule is smart. The retro system is rigged so that the rockets will not fire at all—barring a failure of some kind—unless the capsule is in the correct attitude and aimed in the right direction. The push from the retros could make the capsule start to tumble, and you would have to counteract this tendency immediately with the control system. The three retros should fire in the proper sequence. They are not supposed to go off together. They are timed to fire at five-second intervals, in a sort of rippling effect which provides for a steadier nudge. We could manage to come back if only one of them fired, but this would give us a slow start—and, therefore, a long re-entry and a hot ride. If all three retros failed to fire, however, we could be in real trouble. There would be nothing we could do then but keep going around and around until the capsule lost some of its speed and came down of its own accord. This might not happen for days, and by the time this long voyage home was over the pilot would probably have long since run out of food and oxygen, water and fuel.

Once their duty has been performed, the retro-rockets fall away; the whole package jettisons as the straps which have held them to the capsule are released. Now it is the turn of the smooth, rounded, bottom-side of the spacecraft to do *its* job. It has a vital task, for all of its bluntness.

As we had all learned in our engineering courses, the very act of being hurled up there at such a tremendous velocity will have stoked the capsule full of quite a bit of kinetic energy. By the simple laws of physics, this energy has to be drained off before the capsule can come back down again in peace. The obvious way to dissipate this energy is to convert it into heat and then drain off the heat. The engineers designed the capsule so that the blunt nose would come down first. The nose would sop up much of the friction we were running into and would become quite warm—up to about 3,000 degrees Fahrenheit. But the engineers had already figured out a trick for bringing missile nose cones through the atmosphere without burning them up, and they applied this lesson to the bottom of the capsule. They coated it with an ablative material, a laminated glass resin which evaporates very slowly under the intense heat and peels off a few drops at a time. We call this blunt nose the heatshield, and when we're zooming down in a fireball it is comforting to know that the ablative heatshield is melting away and not the capsule.

While the heatshield is protecting the capsule and keeping it from burning up, there are still a few more fascinating gadgets waiting to play their part. The first of these gadgets goes into action at about 21,000 feet, as the capsule continues to hurtle down. This is a small drogue parachute, six feet in diameter. When it pops out from its cannister on top of the capsule, we are once more a creature of earth, back in the safe embrace of our own environment.

The drogue chute is made of light nylon and is full of open spaces for the air to pass through. It is not supposed to slow down the capsule. It was designed simply to stabilize the capsule until a far more important moment came in the descent. In our original design, the drogue chute was released and opened up as the result of a barometric signal. That is, when the outside pressure reached a certain point, the capsule knew it was at the correct altitude and sent out an electrical impulse to the firing mechanism that deployed the chute. Early in the program, we set the drogue for 45,000 feet; later on we decided this was too high and set the barostat for 21,000. We also rigged the drogue circuit so that if it did not deploy automatically, we could release it ourselves by pushing a button on the instrument panel. Just before my own orbital flight, we altered the system again. The drogue had popped out prematurely on John Glenn's flight and we were afraid it might do this on mine if we left it alone. So we changed the system to allow me to deploy the drogue manually. (We did see to it that the drogue would open automatically at 11,000 feet if for some reason I had failed to get it open earlier than that.) This was another example of our learning as we went along and giving the Astronaut more control over some vital functions of the flight instead of leaving it all to the automatic systems. We were going to be masters of our own destiny to a greater degree on the longer flights into space, and we all felt it was good to start handling some of these affairs on our earlier missions. At any rate, just before the drogue opens up, the capsule would be dropping straight down towards the ocean at a speed of about 600 miles per hour. Then, at 10,000 feet, another barometric signal would be flashed, the drogue would fall away, and the big main chute would take over. This is a ring-sail parachute which is 63 feet in diameter—about three times as large as the parachute most pilots wear—and it is flung out into the air by a mortar that has been fired by electricity. It is built to support a heavy load, and since the capsule and the Astronaut might plunge quite a way into the sea if it failed to open up, we have a spare parachute packed away in the same storage space. The Astronaut also has a ring on the instrument panel which he can pull if the chute fails to come out automatically at 10,000 feet. On my flight, I did just that. I gave the chute exactly 500 feet to pop out after we passed the 10,000 mark, and when it didn't I pulled on the ring. The parachute deployed then in good shape. The capsule has been falling now at a speed of about 200 miles per hour, so we get another pretty good jolt when the main chute opens up and slows us down to about 20 miles per hour. We are subjected to a spike of about four Gs for a fraction of a second when this happens, and we get pinned back in the couch again by this extra weight of our bodies.

I have always thought that using parachutes to come back down after looking at the stars was a rather unglamorous way to end such a glamorous trip. You usually think of using parachutes only to get out of a burning airplane or for some other kind of emergency. Under the circumstances, however, it was the best method the engineers could devise for Project Mercury. If we had tried to glide back on wings, we would have had to go to an entirely different system which would have been much too heavy for the Atlas to lift off the ground and would have taken us far longer to perfect. We knew that parachutes would work because we had been using them for years.

After the main parachute has opened up and we are drifting down at a speed of about 30 feet per second, we have one more good jolt to wait for —the impact with earth. We hope we will land on water because this is where everyone is prepared to pick us up. Even on water, however, the capsule could be slapped around quite a bit if a sudden gust of wind blew up some good waves, and we could get a quick jolt of 30 Gs or so. Some wise engineer thought about this problem and provided us with one more ingenious device which we could literally fall back on. This is a sturdy cushion full of air which fluffs out beneath the capsule soon after the main chute has deployed and is designed to take up a good deal of the shock. The bottom part of the cushion is the round heatshield which has already done its main job by now and has dropped down, tugging behind it a four-foot skirt of rubberized fiber glass which has a series of holes punched in it. Just before impact, the skirt is filled with air and the air is partially trapped inside. When the capsule strikes the water, the skirt slowly collapses under the capsule's weight like the bellows of an accordian and squishes the air out through the holes to cushion the impact.

Now, as we wait for the ships, planes and helicopters of the recovery fleet to find us, the capsule performs one final feat of dexterity and timing. On earlier flights, it scattered out a big cloud of radar chaff to help the search planes spot the area, and dropped a Sofar bomb into the water which set up an underwater echo that the ships could use to locate the area with their sonar gear. We did not use these two recovery aids on my flight. We had learned by experience that we did not really need them, and we could use the extra weight they displaced for other items inside the capsule. We did hang onto a number of other recovery devices, though, and these put on a real good show for me. A light mounted on top of the capsule started flashing automatically when the capsule hit the water to attract the attention of the recovery forces. Our tests showed that this light could be seen for 40 miles on a clear night. A blob of yellow dye spilled out into the water at the same time to make it easier for the search planes to find the capsule in the midst of all that blue water. And three radio beacons went to work sending out signals which the recovery fleet could home in on if it was still having trouble finding us. One of these beacons, which we call Sarah—for "Search and Rescue and Homing"—put out the signal that helped tell the search planes exactly where I was when Aurora 7 overshot its mark at the end of my mission and I had to float around in the water for a while. There was one Sarah beacon aboard the capsule, and I had a smaller version of it in my life raft as part of the emergency kit.

The Mercury capsule is an uncanny machine, really. It was created by some of the smartest people in the world. They took infinite pains with it. The room at McDonnell where the technicians put all these intricate pieces together is kept as sterile as if it were a hospital ward full of premature babies. The room is painted white, like a hospital room, and it has sealed doors and an air-conditioning system to keep the dust out. You have to put on surgical-white smocks, white socks and white plastic shoes before you can enter the room. And the technicians even worry about getting fingerprints on any of the pieces they are assembling because even a tiny amount of moisture might make them rust. Each part that goes into the capsule has had a prototype tested to destruction to make sure it can stand the rough ride and the temperature changes. The test procedures are extremely painstaking. First, one part is tested; then two parts are linked together and both of them are tested as a unit. Then small units are joined into bigger units for further testing, and this process continues until finally the entire machine is ready for a master test. When you realize that one capsule has something like 10,000 parts in it, all told, you can see what a tremendous feat of engineering this is. We even had to figure out a way to put the parts together without using oil to lubricate them—because you do not dare let oil come into contact with the pure oxygen which we would be breathing in the capsule. We would almost certainly have a fire if we tried that.

Then, when it's all put together, it does all of these amazing things. It is a ruthless machine. It never tires. If you make a mistake and tell the capsule to do something that it is not supposed to do, it just won't do it. There is something inexorable about a thing like that. It does just exactly what it is designed to do. If one of the capsules were designed to fail, it would fail. When one of them does fail, however, it is usually because some man has slipped up somewhere and neglected to put in one more backup or to make enough allowances for the fact that his capsule is an absolutely *literal* machine. That is why we have had to spend so much time training with it and getting used to all of its idiosyncrasies. We have had to make sure that we have not given the capsule a single chance to foul us up— not because the capsule itself would necessarily do something wrong, but because it might do something absolutely logical and right one day which we had not thought of and were not prepared for. We had a lesson or two to learn about the explosive hatch, for example. We also had to discover ahead of time that if the periscope does not retract all of the way and close its little door tightly behind it, the heat will rush in during re-entry and could suffocate us. The machine is not foolproof—nothing that man has made can be really foolproof. But it is literal.

Now that we have it, the capsule has proven to be such an uncanny gadget that there are literally thousands of people—including the seven of us—who are huddling over each new model and tracing back through all of the seven miles of wire and all the tubes and circuits trying to find out exactly what it is that we have created. We really do not know yet, for sure. But it is a fascinating device and we feel that it has provided us with a basic concept for manned space travel which we can use, with various adaptations, for years to come. It is also a magnificent thing to operate. If somebody asked me to sit still in an easy chair for six hours at a time, I do not think I could do it. But I do not mind at all lying cramped in that capsule, practicing and practicing to make sure that I can handle it. It thrills me every time I watch it click away and start to work.

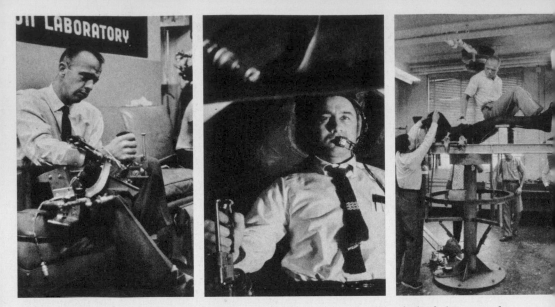

Early in their training, the Astronauts had to get used to the feel of the controls they would use in space. Above, Al Shepard practices moving the 3-axis control handle with which the Astronauts will maintain the capsule's correct attitude during flight. Gus Grissom tries out the same handle. John Glenn gets ready to lie on a contour couch, part of a device known as ALFA ("air lubricated free axis trainer"). Compressed air activates jets which Glenn can use to tilt or turn the couch [OPPOSITE PAGE].

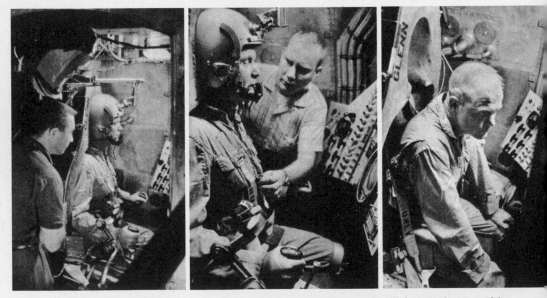

A huge centrifuge trained the Astronauts in coping with the G forces they would encounter during flight. This was a sealed gondola on the end of a 50-foot arm which could spin at high speeds to simulate the sensations they would experience in the capsule. Above, Grissom looks on as Glenn prepares for a test run; Air Force Lt. Col. William Douglas, the Astronauts' doctor, adjusts Glenn's straps. After the ride, the flesh above Glenn's ear is creased by the pressure of his helmet.

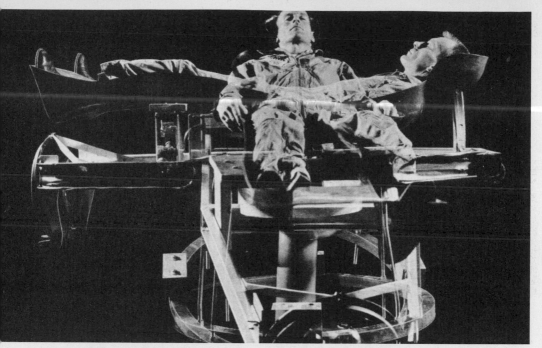

Lying on the ALFA trainer, Deke Slayton activates controls which make the couch spin or tilt through the same yaw, pitch, and roll axes that the capsule will experience during flight. In this double exposure he rides flat (side view), then tilts himself upward. Each Astronaut spent about 50 hours on the trainer to get used to the sensation of controlling his movement and attitude. Through a periscope mounted at their feet, the Astronauts watched the earth's surface moving past them on a screen in order to learn how to maintain a correct position with respect to the earth.

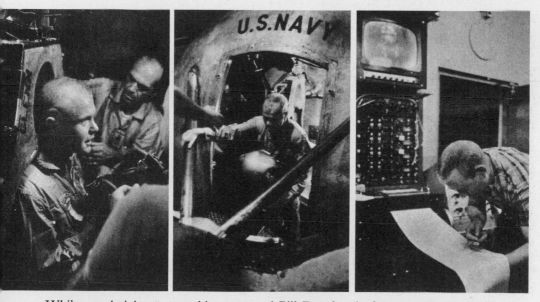

While a technician loosens his straps and Bill Douglas (in foreground) holds his helmet, Glenn describes his ride. He climbs out of the gondola for a debriefing on the simulated mission. Throughout, Douglas had monitored the Astronaut's respiration and heartbeat at a panel inside the centrifuge control room [RIGHT]. Douglas could have stopped the ride in a moment, if it appeared the Astronaut was having any trouble.

EYEBALLS OUT,

Donald K. Slayton

John H. Glenn, Jr.

Malcolm Scott Carpenter

Virgil I. Grissom

Alan B. Shepard, Jr.

EYEBALLS IN

In this chapter, five Astronauts take turns describing the training program which gave them a taste of the rigors of space flight and prepared them to handle the complex machine they were helping to design.

A STANDING START

Donald K. Slayton

SINCE THERE WERE no ground rules for training Astronauts when we started to work, we just plunged in. We knew we had a lot to learn. So when we started chewing into it we adopted three basic principles. First, we would use any training device or method that had even a remote chance of being useful. Second, we would make the training as difficult as possible so that we would be overtrained, if anything, rather than undertrained. And third, except for some wise scheduling of our time, we decided to conduct our training on an informal basis. Everyone assumed from the start that we were mature, well-motivated individuals. Everyone knew we were all eager to make good. So no one had to goad us to get our homework done. The basic idea was to drill us on all of the details until the new art of space flight was as much in our bones as flying had been all along. But we were left pretty much on our own. We knew what subjects we had to learn and what skills we had to master. NASA left it up to us, more or less, how and when we would get it all done. No one gave us exams, as in a school, and no competitive scores were posted. We knew all along, of course, that our individual performances were being watched by NASA's managers and that theoretically any one of us could be washed out of the program if he failed to keep pace with it.

First of all, we started reading all the books and manuals we could get our hands on. Most of us had been only briefly exposed to such sciences as astronomy, geophysics, physiology, meteorology and astrophysics. None of us has had time even now to become real experts in these fields. But even on our first orbital missions we wanted to try to bring back at least a little bit that man had not known before. And in order to know what to look for and where to look for it, we had to know something about what man already knew and how he knew it.

So we sat in our classroom at Langley—where we had seven desks and seven chairs and a blackboard—and listened to hours and hours of lectures. Lieutenant Colonel Bill Douglas, the Air Force doctor who was our advisor on aeromedical matters, brought in the same kind of working models of the human body that the doctors use in medical school and started to teach us, in tremendous detail, about how our circulatory system works, how and why weightlessness would affect our muscles, and how pulling high Gs would affect our breathing. Bill wanted us to understand our own organs and physical reactions so well that when we got up into space we could try to diagnose any troubles we had and either take care of them ourselves, or, at least, know how to discuss them in detail over the radio so the doctors on the ground could give us some good advice while we were zooming along.

The engineers came in and described to us the workings of the boosters that would be taking us into space and the mathematical formulas and theories of flight dynamics that would be used to plot our own trajectories and achieve the precise paths we would follow on our orbital flights. There was a lot of basic physics in all of this, and we had to brush up on our calculus and review the laws that Kepler and Newton had propounded centuries before which still fit the tricks of gravity and force and motion that we would be trying to perform. There was also a lot of new information to sponge up that we had never had to pay much attention to as pilots, because we never strayed much farther from the earth than 50,000 feet up. We had to learn about the mathematics involved in getting us into an elliptical orbit, which the engineers told us would be the safest flight path for us to take and would require the least expenditure of energy on the part of our boosters. We learned about celestial mechanics: how celestial bodies like stars and planets—or our spacecraft—move through space. The astronomers began to fill us in on what they already knew about the universe and described some of the things they hoped we could help them find out. They had already arrived at the weight and distance and composition of many of the stars, and they had measured the temperatures of some of them and speculated on their life histories. But they did not know exactly where our earth fitted into this colossal system, and how it was related to some of the bodies we would have a chance to see clearly for the first time. We knew that some of us might eventually make a landing on the moon if the Mercury system worked out as well as we hoped, and the astronomers were able to tell us what they knew about our nearest neighbor. They pointed out that they did not believe the moon has much of an atmosphere, because the edges of it appear very sharply defined instead of hazy, as they would probably look through a thick atmosphere. And when a star disappears behind the moon, it is cut off sharply, not gradually as would be the case if it slowly disappeared behind an atmosphere first. This phenomenon is known as star occultation, and we planned on our very first orbital flights, once we got beyond the atmosphere and could get a clear look, to watch closely for examples of star occultations and see if they looked the same from up there as they do from down here on earth. We also had to learn something about the propulsion systems we would use to get into space. The ratio between the thrust of the boosters and the specific impulse of the fuels they would burn was the kind of thing we had to think about. We had to understand the thermodynamic problems of re-entry through the atmosphere, and the principles of velocity and guidance in rocketry which would affect our flights. All told, we had the equivalent of about a graduate-school course in astronautics, which is the technology of putting manned and unmanned objects into space and controlling their movement and their destiny.

The people from McDonnell gave us lectures on the spacecraft which they were designing and beginning to put together. This was a brand-new kind of vehicle, and it was not so easy to keep up with it—not even for McDonnell. They put out a manual for us in September 1959, for example, which told us everything they knew about it themselves at the time. But the manual went out of date as fast as the capsule grew—like Topsy, I guess—and they had to put out another one two months later. In the meantime, in order to get our homework done—and take a look at what it was the engineers were talking about—we had to work with some early drawings of the spacecraft that had been included in the original specifications. This was a little bit like learning how to cook from looking into some chef's garbage pail. But we caught up pretty quickly on this subject, especially when we broke away from Langley and started visiting some of the places where the system was being built, like the McDonnell plant in St. Louis, where we got to put our hands on the capsule for the first time and could begin to give the engineers a few of our own ideas before they had gone too far. We went to Akron for fittings of our space suits and to Los Angeles to fit our helmets, and then to Dayton to test our new suits in a heat furnace. At the Convair plant in San Diego, we saw the Atlas under construction and watched some static tests of its engines. It was quite a thrill, seeing all of that power that we would be sitting on top of. We went to Cape Canaveral to become familiar with the layout there—though it would be a long time before any of us would start using the launching pads. We wanted to see for ourselves what a booster launching was like so we could keep in mind some of the situations we might run into later before we got started on our training. But mostly we wanted to get acquainted with the launch crews we would be working with and discuss our mutual problems with them. We met a tremendous number of people in those first few months, and we were impressed by the caliber and dedication of the personnel we would be knowing so well over the next few years. According to our records, we spent about one day out of every three traveling during one four-month period, and each one of us piled up about 20,000 miles on the road. That seems like a lot of moving around. But, looking back, I'd say that it was worth every mile of it. It was a good cram course on what the country was doing for the space effort, and I think we probably came home more inspired to catch up with it. I guess we asked a lot of questions. We often asked a lecturer to go on talking overtime so we could be sure we understood what he was trying to tell us before he got away and they moved us on to something else. A number of these questions paid off. I remember one day during a medical lecture when John Glenn asked Bill Douglas if he thought the restraining harness that was designed for us might not leave us open for some strong negative Gs by allowing the Astronaut's abdomen to be thrown forward too far at a certain point in the flight. Bill said it probably would, if the force was strong enough. So we made a note to start fighting for some revisions on the harness.

We also had to learn about our new machine. That was really the most important thing. All of us knew quite a lot about the vagaries of untested airplanes. But we knew practically nothing about missiles and rockets when we first arrived at Langley. We took it for granted that boosters could explode on the launching pad and that failures up in the air were everyday risks in the missile business. We also knew, after a lecture or two from Convair, that the big Atlas had about 40,000 parts in it, any one of which could conceivably go haywire at any second, even after you thought you were on your way. The spacecraft had several thousand more parts in it that could go wrong. But we also knew that any part in your automobile could act up—like the brakes when you are in the middle of an icy hill. We figured that the Atlas machine would work smoothly whenever we climbed in and stepped on the starter—provided the people who built it and aimed it knew what they were doing. Part of *our* job was to satisfy ourselves that they did.

The rest of our job was to satisfy ourselves and everyone else in the picture that we could handle the machine once they gave it to us. Here was where our flying experience came in, and our ability to make fast, correct decisions in the clutch. We knew we had to practice and practice until we were sure that we could not only stay on top of any specific emergency that might come up on a flight, but that we could also handle troubles when they came in bunches. You had to train for the extreme situations in which two or three troubles could pile up on you at once, like having to fire the retro-rockets manually when some of your instruments had just gone out and you also had just lost cabin pressure, which meant that your suit would suddenly inflate automatically and limit your ability to move your arms freely. A multiple electrical failure could cause a crisis like this. Without electricity, you would have no fans to move the stale oxygen, no autopilot to control the attitude, no air conditioner, no flight instruments. We figured that if we could train ourselves well enough to handle a situation like this, we could handle anything.

Some time before we were scheduled to fly actual missions, McDonnell built a pair of realistic trainers for us to try things out in. We called these our procedures trainers, and they were exact replicas of the spacecraft we would ride in. They had the seven miles of wire in them, a complete oxygen system and all the rest, and we used them to simulate all of the routine and emergency situations that we thought we would ever run into on a real mission. One of these trainers was installed at Cape Canaveral so that the same people who would monitor our flights could also get some practice, along with us, on how to control the mission. You could preset the trainer for any kind of mission you wanted to fly, and you could crank emergencies into the flight plan so that the electronic computer which operated the trainer would throw these at you automatically at the proper time. Then you would go on to figure out a way to solve the problem without tumbling out of control or burning up or using up too much fuel. You were sitting on the ground in the procedures trainers, of course, and the only way you could get hurt would be to snag your fingernail on a switch or something. But you could damage your ego pretty badly. And you could certainly foul up your chances of ever flying one of these gadgets if you could not prove that you understood it on the ground and had it tamed.

Then, when we were about ready to start flying real missions, we worked out some of the details of our actual flight plans on the trainers to give us an exact feeling for the timing of all of the events and to help us spot periods of time here and there when we could schedule various experiments that we wanted to carry out on the flight. The trainers came in very handy. But they were no substitute for the real thing. No simulator ever is.

The idea of using a simulator for training was not new, of course. Soldiers learn how to aim a rifle by pointing wooden sticks at a target and squinting across the splinters. Military pilots are trained to fly complex modern aircraft with weapons-systems simulators. Roman soldiers had to toughen up by throwing extra-heavy spears—so the ones they used in combat would seem light by comparison.

We used more or less the same principle in Project Mercury. Our training advisors, who were led by a former Navy psychologist, Dr. Robert Voas, decided to make things a bit harder and tougher for us than they probably would be in actual flight. They had us pull more Gs on the centrifuge, for example, than they estimated we would ever encounter on a mission. We tumbled and twisted on the MASTIF trainer at a rate that was quite a bit higher than anything we expected to run into on a real flight. We deliberately overstressed and overstrained ourselves every chance we had. There were several reasons for this. One was that our advisors wanted to toughen us up. A pilot can build up a certain amount of tolerance for G forces, for example, if he is exposed to them often enough. Another reason for being tough was that you just cannot simulate space travel on the ground. You can simulate isolated parts of it—like G forces or disorientation or excess CO_2 production or attitude control problems or how to get yourself out of a tumbling situation. But no one has found a way on earth to simulate the sensation of weightlessness for longer than one minute at a time—which is not long enough to give you much training. And you have to accomplish even that in the back seat of a jet fighter plane as it comes out of a fast climb. You obviously cannot put the entire simulator in the back seat of a jet in order to combine this experience with other factors like mission control at the same time. Therefore, to make sure that we were strong enough and well enough prepared to do *all* of these things together, we accentuated the training on each one of them and then hoped that when we put them all together for the first time in space the total experience of all of the simulations we had had would add up to equal the stresses we would meet in actual space flight. We could not be sure, of course. When you are able to deal with only one or two problems at a time, and not the whole spectrum, you cannot conclude that you are adequately trained to solve them all at once. That would be up to the man, once he got up there. And that is why Bob Voas and his staff left it largely up to us to build ourselves up until we were sure we were ready. Our advisors tried to help us by making all the stresses and strains available on the ground, and by showing us how to make the best use of them.

THE WHEEL

John H. Glenn, Jr.

I WOULD SAY that the large centrifuge was the most spectacular of all the stress-making machines we have used in our training. The "Wheel," as we called it, was a standard piece of equipment at the Navy's Acceleration Laboratory in Johnsville, Pennsylvania, and it combined many of the control problems we would encounter in real space flight. It is the most advanced machine of its kind in the world, so far as we know, and we started using it almost from the beginning. It consists of a pill-shaped gondola, 6 feet by 10 feet, suspended at the end of a 50-foot arm. The arm and the gondola are spun around by a 180-ton, 4,000-horsepower electric motor that is capable of generating enormous speeds. We were whirled around in it to duplicate the stress of acceleration, or G forces, that we would be experiencing at various predictable points during a space flight. These would come at the very start of a flight, for example, when we were being boosted up at a tremendous speed against the pull of gravity. Acceleration occurs when a moving object gains speed or when its direction or movement is changed. You experience the same sensation when you turn a corner in a fast-moving automobile and feel your body being pushed away from the direction of the turn. If the car were moving fast enough, you might even feel as if a weight were pressing you against the side of the car. At each moment of the turn, the car is taking your body in a new direction of movement, but your body tends to keep going in the direction in which it was moving the moment before. Magnified a good number of times, this is the same kind of sensation that the Astronaut feels when the spacecraft is lifting him up—except that in this case he is being pushed back into his couch instead of to the side. This force of acceleration is measured in Gs. One G equals the normal gravitational pull of the earth. As you read this, you are at 1 G. You are so accustomed to the feeling that you do not notice it. But you would feel the difference if you started pulling more than that. You would feel heavier, for one thing. For every extra G of acceleration force, your body is subjected to a stress that is a multiple of your own weight. For example, a man who weighs 200 pounds and is experiencing 3 Gs will be subjected to a force of 600 pounds. When you get up into much higher Gs, a man could easily become unconscious as a result of the extra force exerted against his body. That is why we practiced over and over on the centrifuge. It taught us how to fight off the forces and keep our bodies and our brains working even under tremendous stress. We found that we were able to carry out certain duties without much trouble up to about 14 Gs. We did not expect to experience such high forces, but we wanted to test ourselves under these conditions in order to give ourselves a margin of safety. We already knew, from data that had been collected on other rocket flights, the kind of forces that we would be experiencing in the spacecraft. We estimated that we would experience a peak force of about 8 Gs. A flight plan or "profile" of the missions we planned to fly on both the Redstone and Atlas boosters was fed into a computer, and then the computer issued appropriate instructions to the motor of the centrifuge which drove the gondola we were riding in at the exact speed and intervals of time we needed to simulate the real thing. On a typical run in the centrifuge, we would climb into the gondola with our helmet and, in the most strenuous tests, our full pressure suit on, and let the technicians strap us into our molded couch. Part of the air was evacuated from the sealed gondola to simulate conditions we would experience in space and permit us to experiment with the emergency use of our pressure suits. The air we breathed during a ride was produced by the same closed-circuit environmental control system that we would use on a real flight. We had an instrument panel and controls similar to the ones we would have inside the spacecraft, and we handled it the same way we would control a Mercury capsule in flight. Before each run on the centrifuge, the medics would tape electrodes and a respiration gauge to your body so they could keep a continuous record of your heart rate and breathing rate during the ride. Then the gondola was turned so that you faced inward to let the G forces push towards the front of your body, from front to back, and shoving you into the couch just as you would be pushed back by acceleration during launch. We called this riding "eyeballs in," which meant that if our eyeballs were actually free to move that much—which they were not, of course—we would feel them being pushed clear back into their sockets. We also made runs to simulate the forces of deceleration, or negative Gs, which we would experience if we had to abort the mission just as we reached the point soon after launch where the highest aerodynamic forces were being concentrated against the capsule. You would be riding up against a force of 982 pounds per square foot. If you suddenly had to abort and come to a rapid stop under these conditions, the G forces would reverse, and you would feel them going the other way. They would push you from back to front and would tend to lift you out of your couch instead of pinning you in it, and throw you against your restraining harness. Since your eyes would tend to pop *out* of their sockets on a ride like that, we called this riding "eyeballs out." At one point in our training we took a few rides on the centrifuge that called for a particularly wrenching combination of going "eyeballs in" and then "eyeballs out." This was to prepare us for the rather slim possibility that we might have a booster failure early in the flight while we were still on top of it and would be pulled away from the booster by our escape rocket at a high rate of acceleration. The capsule would do some tumbling then as the atmosphere slowed it down, and we would be subjected to a wide range of strong G forces in a very short space of time. If we were pulling 8 Gs as we went up, for example, and suddenly got a reverse force of 8 Gs, we would experience a total range of 16 Gs. We simulated this on the centrifuge by flipping the gondola over in the middle of a run and going in two seconds from "eyeballs in" to "eyeballs out." These were rather strenuous runs, as you can imagine.

To get back to a typical ride, the centrifuge starts up and the Gs begin to build slowly. Almost immediately, the pressure forces you back against the couch. To simulate the booster's acceleration in the early stages of a flight, the forces climb to 4, 5, 6, and 7 Gs. When you approach 8 Gs, you can feel quite a pressure on your chest cavity and you have to work at it to breathe. Some of us found it easier to breathe if we grunted the air out. I personally preferred to hold my breath, let it all out, then hold it again. We also had to strain all the muscles we could—our legs, calves and arms—to keep the blood flowing away from the abdomen, where it tended to pool under high G forces, and headed back towards the heart to keep it working. The centrifuge is a tough machine and you have to fight against it without any letup. If you relax, the blood begins to drain from your brain, your vision begins to blur in a sort of "gray-out," and you are experiencing the first steps towards unconsciousness. After a few seconds at a peak of 8 Gs—which is as high as we predicted we would go on a normal launch—the computer feeds the correct instructions to the motor on the centrifuge and you can feel the acceleration forces begin to taper off. On the centrifuge, which cannot simulate weightlessness, they decrease gradually to the normal 1 G, and then start rising again to simulate the period of retro-firing and re-entry into the atmosphere. They reach a peak of 8 Gs again at the theoretical altitude of 65 miles or so, and then drop off rapidly to the point on the flight plan where the drogue chute would open up on an actual flight to stabilize the descent, and we feel a quick surge of Gs again.

We made some interesting discoveries during the centrifuge runs that helped us on the program as a whole. Someone suggested, for example, that we try working the "eyeballs out" runs and some of the tumble runs without the head restraint that we thought we would need in the capsule to help hold our heads in place while we were pulling high Gs. We tried it without the restraint and it worked. The space suit itself was such a tight fit that it acted as a head restraint all by itself. It was built so that a man could not bend his neck or get his head down if he wanted to. Taking out the head restraint saved some weight in the capsule, and this was important. You never put anything into a flying machine that you don't need.

Some of these practice sessions were very strenuous. They varied in length according to whatever phase of a space flight we were simulating on the wheel—launch, insertion into orbit, retro-firing or re-entry. After a number of runs on any one day, you knew you had done some work. We got used to the sensations as time went on, however. I do not believe that a person actually builds up a physical tolerance to G forces, but you do develop various techniques for coping with them and you get more or less inured to the stresses they put on you. It was uncomfortable work sometimes, especially if the couch happened to rub against a particular pressure point on your body as the wheel whirled you around, and you could feel a little sore for a while afterwards. We were all in good physical shape before we started taking the rides, however. We were exercising regularly and keeping ourselves in shape for the training. Also, Bill Douglas, our flight surgeon, checked us over carefully before and after each trip. If we were suffering from a slight cold, Bill would not let us go. If we developed any signs of fatigue during a run, he would not let us continue. As soon as a run was over, he would climb into the gondola before we'd been unstrapped and look us over to make sure we didn't show any ill effects. The medics took blood and urine samples from us after many of the rides and examined them for any abnormalities that might indicate the stresses were too great. Bill usually monitored the controls while the centrifuge was operating, and he had graphs in front of him which showed our respiration and heart rate. A closed-circuit TV camera mounted inside the gondola let him watch our faces for signs of strain or unconsciousness. If Bill had any indication of trouble he could flip a switch to bring the gondola to a fast stop.

In addition to all this, Bill took runs on the centrifuge ahead of us so he would know in advance what we were going through. I call that a real fine bedside manner. Along with Bill Augerson, another of the doctors who helped monitor our progress and give us advice along the way, Bill Douglas made a number of runs on the centrifuge to simulate what it would be like if the capsule started to tumble in space and expose us to a quick succession of high positive and negative Gs. Your head snaps forward on a run like this, and your legs tend to fly out of the couch. Both Augerson and Douglas made these runs at various rates of rotation and at various angles to see what the effect would be when the capsule was tumbling at various speeds and was in various specific attitudes when the tumbling began. They tried everything out in both pressurized and unpressurized suits and with the pressure inside the gondola varying all the

way from sea level to the five pounds per square inch we would normally
have inside the capsule. Out of all this, we got some good advice. We had
experienced some pain on the higher G runs—up around 16—and Bill
Douglas suggested that we grunt and squeeze out words to relieve the
pain. He told us this procedure would help pump blood from the abdomen, where it was pooling, and back into the heart by changing the pressure in the thorax. The change in pressure that you get as a result of
grunting acts as a pump itself, and this relieves the heart of some extra
work. The technique worked fine and it made some of our runs on the
wheel a good deal easier.

*Lieutenant Colonel William Douglas, a handsome, soft-spoken Air
Force flight surgeon with prematurely graying hair and a penchant for
studying the stars with his own telescope, had what was undoubtedly the
most unique practice in U.S. medicine. He not only guided his seven prized
patients through the physical perils of training and actual flight, but he
also served as friend, confidant and personal physician to the Astronauts
and their families. He did, indeed, have the perfect bedside manner.*

*"I went through the tests with the boys," he explained, "so I could
speak their language—and so they could never tell me I didn't know what
I was talking about. I took the full Redstone profile—up to 11 Gs. And
I made the other runs that we knew might make them dizzy. I figured
that if I could tell them whenever they got into trouble that* this *is how
your physical mechanism works and* this *has helped* me, *they might take
my advice and benefit from it. They didn't always take it. Some of them
grunted to help them breathe and some did not. They all figured out
their own capacity, and it was not always the same. Here again, they
were all different. There were no two of them alike."*

THE ONE UNKNOWN

Malcolm Scott Carpenter

THERE WAS REALLY only one big unknown in the entire program. That
was how our bodies would react to the long periods of weightlessness we
would be subjected to on the orbital flights. John Glenn has just described
the effects of high G forces we would encounter on the way up and on
the way back down again. Weightlessness is just the opposite. It is the

state of zero G—of no G forces or weight at all—which we would experience the moment we went into orbit, when the centrifugal force which the booster's tremendous velocity had imparted to us would precisely balance out the pull of gravity that we had lived with all our lives. We knew almost nothing about what effects this strange condition would have on an Astronaut when the program began. We knew that some of the doctors had warned that prolonged weightlessness *could* have some dire effects. It might affect our respiration or the heart, for example, or the stomach might stop digesting food if there was no pull of gravity on the muscles, or the circulatory system might collapse entirely for the same reason. Some doctors suggested that the sensory organs of the middle ear, which normally tell a man which end is up, might cease to function properly and that we could get badly disoriented. We also heard about a test that a young Air Force doctor conducted on himself at the School of Aviation Medicine. He was dressed in a skindiver's suit and strapped to an aluminum beach chair, and he spent an entire week floating in a 400-gallon tank of water. Only his head was above water; his body, which was supported by the water, was almost in a state of zero gravity. When the week was over and the doctor got out of the tank, he discovered that his muscles had become soft, his circulation was poor and his bones had apparently lost some of their calcium. He also noticed that his mental response to simple efficiency tests had deteriorated during the week and that he had grown more confused with each day he remained in the tank. This test was not conclusive, but it did indicate that some precautions would have to be taken in the future on long space flights. There was no need to worry about the relatively brief flights which we would make early in the program, however—including the four-and-a-half-hour orbital missions. In fact, after going through several tests of weightlessness on the ground, I thought it was more delightful than dangerous. Compared to some of the other problems we knew we would face, it seemed like a real piece of cake.

Dr. William Douglas was aware of the dire forebodings of some of the other experts on the subject of weightlessness. But he personally was unworried about this particular hurdle his patients would have to jump, as he was about most of the other problems. "I don't suspect that weightlessness over any length of time will be a serious problem or make people sick," he assured them. "Before Titov went on his trip and reported that he had experienced a little trouble I would have said, 'Hell's bells; no problem.' Now, since we know a little bit more about the trouble he had, I'd just leave off the 'Hell's bells.' There will probably be a certain amount of deterioration of the muscles and of the muscle tone, because a man who is floating around with no force of gravity pulling on him every second will discover that his muscles have very little to do any more. They might even atrophy a little over prolonged periods. But that's no great problem, either. The man can do exercises while he's up there and keep his muscles alive and working."

Even though it is impossible to simulate weightlessness on earth for longer than a minute at a time, we launched into a full-fledged cycle of training on the subject so we could all become as familiar as possible with the sensations involved. The tests we took also contributed to a detailed study of the subject which was being conducted by the Air Force School of Aviation Medicine. We all took turns flying at Edwards Air Force Base in California, where we went up in the rear cockpit of a two-seater F-100F jet trainer. Air Force pilots took over the controls on some of the flights so we could concentrate on our reactions and run off a few experiments. We had a simple and colorful gimmick to tell us how we were doing—a bright orange golf ball.

The trick was to forget the instruments and keep your eye on the golf ball. It dangled on a short nylon string right in front of you, and so long as you could keep a slack loop in the string and keep the ball hanging right in front of your face, you were doing fine and you knew that you, the plane and the golf ball were all weightless. We could climb up to 40,000 feet at 600 miles an hour, then nose over with the jet afterburner on into a 30-degree dive. At 28,000 feet, where we were going 900 miles an hour—a speed of Mach 1.3—we suddenly pulled through and started to zoom up again at a 55-degree angle. This maneuver converted some of our speed into zoom energy. Then, just at the right moment, the pilot pushed the stick forward slightly. If everything had gone just right, the inertial force of the climb would exactly balance the pull of gravity. The golf ball would dangle in front of us and we would be weightless.

It was a very pleasant sensation, actually. We had some work to do, of course, to pay for the ride. An instrument panel with red buttons and orange lights was rigged up in front of you. An orange light would wink on just as you were being forced up out of your seat during the weightless period—or back *down* in your seat during the pullouts at the end of each parabola—and you were supposed to push the button nearest to it to turn it off. The idea was to keep punching out the lights as fast as possible. One would come on just as you turned another one off. I noticed that it took a few seconds to get used to the sudden changes in Gs when we went from weightlessness to a 3-G pullout. But, all in all, the test was a breeze. Since my arms were not burdened by the usual weight when I moved them, I tended to overshoot just a little when I reached for the buttons and missed hitting them the first time or two. During the pullouts I had a tendency to reach beneath them. But I worked out a system, using my lap and legs to steady my arms, and got so I could work out a rhythm of operation and handle the lights very well. I decided on the basis of this experience that I could do the same sort of thing in the capsule and that weightlessness should not affect our speed or efficiency on any of the Mercury flights.

We also practiced eating and drinking to see if we could perform these functions in a weightless stage. The cuisine was terrible. We had puréed beef, some fruit and lukewarm tomato juice, orange juice and water. All of this tasteless but nourishing pap was stored in little tubes or squeeze bottles like those plastic catsup containers you see in hamburger joints. We had to open up the faceplate on our helmets and squirt the food into our mouths as the plane arched over on its long parabola and became weightless for about 60 seconds. We worked out a variation or two on this theme. Al Shepard held the bottle a sporting distance in front of his mouth and really squirted the food in. Gus let some orange juice drift out of his bottle and enjoyed watching big bubbles of juice float around the cockpit with him. Deke gave his bottle a squeeze test before he left the ground and nearly drowned himself in tomato juice.

It was not nearly as difficult performing these tests as it was to make the airplane go weightless when it was our turn to take over the controls. It was a real struggle for me to stagger through the long parabola, and I found it was almost impossible to keep that ridiculous little golf ball from bouncing all over the cockpit. It took a very delicate juggling of the stick and power to make it work. If I used too much power, the ball swung back and almost banged me in the face. If I held too little power, or gave too much forward pressure on the stick, the ball swung forward and glanced off the canopy. I could imagine what the other pilot in the front cockpit was thinking. "This character's an Astronaut?" he probably asked himself. "Why, he can't even fly!"

The doctors had us rigged up with three electrocardiogram leads which were taped to our bodies, and these fed our heart tracings, blood pressure and respiration rates to the ground over a telemetry link that let the doctors watch our progress and keep a check on our reactions. I felt absolutely no apprehension at all on any of the flights. There were straps holding me down, just as there would be in the capsule, to give me a sense of security and provide a reference point so I would not become disoriented. Without the normal pressure of 1 G on the seat of my pants, I actually felt elated and free. On one of the flights I was asked to read from a piece of paper while we were at zero G, to find out how our speech would be affected and whether we would have trouble reporting to the ground from a capsule while we were weightless. My voice did seem to go a little higher than normal, but I was able to read without any difficulty. The message itself was very interesting to me. It was much like the reports we would be radioing back one day from an orbiting capsule. It was a routine rundown of technical information and included the position of the capsule, the time remaining before retro-firing and re-entry, the amount of oxygen still on hand in both the normal and emergency systems, the amount of hydrogen peroxide left in the capsule fuel tank for attitude control, and the concentration of carbon dioxide and water vapor in the cabin and in the pressure suit. Reading all of this aloud into a microphone in an airplane high above the California desert seemed so realistic that it was a thrilling experience for me.

In addition to the F-100 flights, which allowed us to simulate weightlessness for about a minute, we also flew a number of maneuvers in KC-135 jet transports which subjected us to 35 seconds of weightlessness, and in C-131 propeller-driven transports which were capable of simulating zero G for 15 seconds at a time. All told, we flew 108 weightless missions and logged a total of some 38 minutes of weightless flight. Some of the reactions we had were rather amusing previews of the kind of skills man must learn in order to maneuver himself in the vacuum of space.

On one weightless flight in the transport plane, for example, I was handed a screwdriver and told to try tightening a screw on a panel. I floated easily over to the screw, inserted the blade of the screwdriver and twisted away. But instead of turning the screw, my weightless body took the twisting force and I went corkscrewing off towards the opposite side of the plane. A little later, when they handed me a wrench to use, I found myself spinning around in the air instead of turning the bolt. After this experience, I had the distinct feeling that weightlessness, instead of becoming a vast problem for us, would be the most exhilarating part of space flight. We might have to work out our in-flight maintenance procedures a little differently in order to keep the nuts and bolts tight during a long flight, but this perfect freedom from the restraint of gravity appealed to me as the kind of element in which I could easily belong. I was certainly right about that. The greatest thrill I had on my own flight in Aurora 7 was experiencing the state of weightlessness. I don't think anyone can imagine until he's tried it what it is like to have no feeling of up or down, to be able to spin a camera around in front of you without its falling. A change of attitude means nothing in this state. Nothing rises or falls. "Up" loses all significance. You can assign your own "up" and put it anywhere—towards the ground, towards the horizon or on a line between two stars—and it is perfectly satisfactory.

THE THREE-WAY SPIN

Virgil I. Grissom

WE MOVED SO FAST from one thing to another that in some cases we outran the people who were building our trainers. One of the most important tricks we had to learn from the ground up was how to manipulate the new three-axis control stick and make the precise adjustments in yaw, pitch and roll that would keep our spacecraft in its proper attitude. The first training device we got for practicing these maneuvers was a crude lash-up which the engineers put together for us in the summer of 1959. It consisted of nothing more complicated than an old-fashioned computer, a panel with a few instruments stuck on it, and a control stick. The computer was replaced a little later by a more sophisticated gadget which somebody cannibalized from the gunnery system of an F-100F jet. We could practice only the fundamentals with this machine. But eventually we got the device we had been waiting for and buckled down.

This rig consisted of our contour couch mounted on a frame which floated on a cushion of air to cut down friction and let it move freely. Our three-axis control stick was attached to it. But, instead of the hydrogen peroxide nozzles which we would use in space to adjust our attitude, the controls worked a series of compressed air jets which tipped and tilted and turned the couch on the same roll, pitch and yaw axes that we would maneuver along in the spacecraft. Because of the air cushion, our training directors called this machine the "Air Bearing Orbital Attitude Simulator." This proved to be such a tongue-twister that they changed it to "Air Lubricated Free Axis Trainer," which at least gave us a handy nickname for it—"ALFA." ALFA was set up in a dark room where we could watch a picture of the earth's surface moving past us on a big screen to orient ourselves to the ground. We looked at the picture through a periscope, just as we would in the spacecraft, and we got a very realistic image of the earth passing by as we tilted and turned ourselves around to keep on the right plane. As you trained on the ALFA, you would lie flat on the couch and look down through your feet at a large lens which was connected to the periscope. The main thing was to learn how to manipulate your hand just enough to make the little jets move you where you wanted to go and then keep you there. We each spent about 50 hours on ALFA, getting the feel of it.

Another machine, which gave us our wildest ride, was called "MASTIF" (for "Multiple Axis Space Test Inertia Facility"). It was set up in an old wind tunnel in Cleveland, and we had to go there to make our rides. MASTIF was a huge contraption—almost as large as its full name—and it consisted of three separate frameworks of tubular aluminum set one inside the other. The outside frame was mounted on gimbals or swivels so that it could pitch over and over, like a barrel tumbling end over end. Inside it was another gimbaled cage which spun around and around like a pinwheel. Inside of *that* was another cage, which was shaped somewhat like the Mercury capsule, and would spin around on its own axis like a spiraling football. The purpose of MASTIF was to get us revolving and tumbling and spinning on a dizzy ride in all three directions at once in order to see how well we could manipulate the three-axis controls to correct the fantastic combination of yawing, pitching and rolling which the machine could throw at us.

We knew, from previous tests of our flight system, that there was one brief moment on a real space ride when we might have to contend with an identical situation. It could be hair-raising. This would come on an Atlas mission when the spacecraft was being separated from the booster and was being tossed into space. We would break loose as three small posigrade rockets fired to force us apart from the booster. As they fired, they would build up pressure in the enclosed space between the capsule and the Atlas. This meant that the capsule would come shooting off with an uneven motion, just like the cork in a toy gun. We would have to correct for this "popgun" effect immediately or we could be in serious trouble and might get completely out of control. The automatic pilot is designed to notice this kind of problem and use the small control nozzles on the outside of the capsule to correct it. But if the autopilot failed to work, we would have to take over with the manual controls and operate the jets ourselves. We knew it would be a wild sensation and that we would need a lot of practice to be able to handle it. The MASTIF was built to give us this practice.

A mission on MASTIF began like a carnival ride. You tumbled slowly, twisted and rolled as your body lurched against the tight harness that strapped you to the couch. Then you rotated faster and faster until finally you were spinning violently in three different directions at once —head over heels, round and round as if you were on a merry-go-round, and sideways as if your arms and legs were tied to the spokes of a wheel. It was a wild and sickening sensation. Your vision blurred. Your forehead broke out into a cold, clammy sweat. And unless you could stop all of this with your stick you could get sick enough to vomit. You would not be able to help it.

When it was time to start the big machine, Joe Algranti, the NASA test pilot who had spent more time in it than anyone else, would speak to you quietly on the interphone from his control panel.

"O.K., Gus," he would say. "We are going to run you up to 30 rpm this time. On all three axes."

By now you would have already taken several rides, but on only one axis at a time. First you had a series of spins in the innermost cage, and Joe took you up to 50 revolutions per minute on one of these. This ride was not so bad, except for the feeling that your head and feet were getting too much blood and that it was all draining from your stomach. Then, you tried the outermost cage and it pitched you head over heels at speeds up to 30 revolutions per minute. This was not pleasant, either. Your body felt like a tank full of loose parts and sloshing chemicals. Your liver and stomach rose and fell 30 times a minute. Finally you took a 30-revolutions-per-minute ride in the third and last direction, and this twirled you around as if you were pirouetting on ice skates. None of these rides was so bad by itself. But now you faced a taste of what it would feel like to go all three ways at once.

Bob Miller, who was a NASA project engineer, would throw a set of switches at the control panel, and a loud explosive sound suddenly filled the wind tunnel. Tanks of nitrogen spurted out streams of gas under high pressure to set the machine in motion. It was perfectly balanced, so it picked up speed very quickly. You were strapped into your couch, and your helmet was secured by an elastic cord. The instrument panel was two feet in front of you—just as it would be in the real capsule —and you tried to focus your eyes on a round dial in the middle of it. There were three needles on the dial—for pitch, roll and yaw—and they edged slowly away from dead center to show you the speed and the direction in which you were rotating. You held the control stick in your right hand. And you tried to twist and pull and tilt it at exactly the right moment and with exactly the right pressure to compensate for whatever it was that MASTIF was doing to you. Algranti was calmly reading off the velocity. "Ten . . . twenty." And he asked you if you felt the nystagmus yet. This is the medical term for the particular kind of dizziness MASTIF causes. You felt it just as he asked about it. The instrument panel suddenly blurred, and the dials disappeared in a swirling mass. Bill Douglas had already briefed us about this. The doctors call this sensation "vestibular nystagmus," and it is an uncontrollable movement of the eyeballs which occurs when the balance mechanism in your body gets all messed up from the twisting and tumbling. The sensation is similar to what you feel when you try to count fence posts from a swiftly moving car. All you can see is a sickening blur.

As 30 revolutions per minute, all the organs of your body seemed to be sloshing in all directions. Your head and feet felt full of blood. The cold sweat was breaking out on your forehead. And you had only a very short time to work before you would be violently ill. But suddenly the speed felt constant. Your balance mechanism had adapted itself very quickly. Your eyes shook off the nystagmus, and you began to see the pitch, roll and yaw needles clearly again. They showed that you were pitching head over heels, rolling to the right and yawing to the left. To correct all this you had to tug the control stick straight back, twist it to the right and tilt it to the left—all at once. The needles slowly started to drop towards zero. But now the nystagmus hit you again. The sudden deceleration that came as you slowed down had the same effect on your sense of balance as acceleration had. So once again your eyeballs started moving involuntarily. You realized now that you had to recover from this multiple spin in easy stages, that you could not do it all at once. So you brought the control stick back to neutral, and the eye movement stopped. Then once again you started twisting, pulling and tilting the stick. Little by little, with the nystagmus dogging you all the way, you brought this twisting, turning, rolling machine to a halt. You were glad to get out, because you were sure that just a few more seconds of it would have made you sick. And you had the feeling that if you moved around much right after a ride, someone would have to fetch a mop. The best thing to do was just to stand steady for a minute or two. You even felt like getting sick if you just stood there and watched another Astronaut take his turn. It was as hard to watch all that rotation from outside MASTIF as it was to experience it yourself from the inside.

Even so, I think MASTIF gave the seven of us more confidence about our ability to pilot a spacecraft than any other test we took. The scientists told us that the tumbling of the Mercury capsule would be only one-tenth as violent as the 30 revolutions per minute we suffered through in MASTIF. So if we could toss this spear, I guess we could toss any spear they handed us. We all knew that the best way to face any unknown was to find out all we could about it in advance. You can never have too much training.

Bill Douglas also went for rides on MASTIF to get the feel of it, and he was able to reassure his seven patients that they need not feel embarrassed by the urge to vomit. "There are three different sensations involved in motion sickness on an aircraft or spacecraft of any kind," he told them. "First, there is the pure vestibular sensation you have on MASTIF. This is plain, old-fashioned vertigo. You get extremely dizzy from all that spinning. But it is not the sensation of tumbling and spinning in itself that makes you sick. It's the rapid changes in the rate of motion, the speeding up and slowing down. If you accelerate or decelerate on the vertical axis of your body, the whole world seems to be turning in the opposite direction when you stop. The blurred vision results from your eyes flickering faster than usual as if they were trying to keep up with some object. On MASTIF, your eyes flicker so rapidly that the vision blurs on the horizontal plane, and objects that are standing on the vertical plane get blurred. When you are spinning head over heels, the horizontal line becomes blurred, and you would have trouble standing up if you got out of the cockpit immediately. But your muscular sensations tell you which way is up. And by the time you're unstrapped, the sickness is really over. The second kind of motion sickness comes when you go up and down, up and down, as you do when you fly the zero G parabolas and follow the golf ball. There is no real disorientation here, no vertigo. You still have visual clues—the top of the canopy and the bottom of the cockpit—to keep you oriented. We think the cause of nausea here is that your gut is tugging away at your intestines from all that rollercoaster motion. If a surgeon who has cut into a patient should grab for the intestines and tug away at them, he could make the patient vomit. Normal seasickness and airsickness is a combination of these two sensations. But it's mostly the first, the vestibular reaction or the dizziness, that affects you. The third factor, which we call the Coriolis effect, comes from going around in circles. This will make anybody sick, and this was probably what bothered the Russian Cosmonaut, Titov, when he was in orbit. Perhaps the Russians figured out a way to make Titov roll or spin on purpose, to help stabilize his capsule the same way spinning stabilizes a gyroscope. They did mention rotating the capsule to help equalize the heat. We understand that he reported he felt better when he made a determined effort to hold his head still. That would have helped. For when you are rolling, everything gets blurred because your eyeballs are twisting on you. This can give you an almost incredible feeling of vertigo. And it can make you feel like vomiting, whether you ever do or not."

WHAT TO DO UNTIL THE SHIP COMES

Alan B. Shepard, Jr.

WE SPENT TWO YEARS doing many things and following up many avenues to make sure we had not overlooked anything. We crammed ourselves full of knowledge. We built up our stamina on the big machines. And we got thoroughly familiar with the spacecraft that we would fly. Some of this was fairly exotic stuff. For we were preparing to penetrate an environment that no one had ever dealt with before. Some of it, however, was just plain down-to-earth hard work.

It would be a bitter irony if this most exciting of voyages through one hostile environment which we did not know too well were to end with the Astronaut getting lost in the midst of another environment which he knew so much better—the earth and the sea. We knew this could happen. If the retro-rockets failed to work on schedule, or if the Astronaut had to declare an emergency and bring himself down over an area where we were not so well prepared to pick him up in a hurry, he might have to fend for himself for a while. So we all learned how to do this. The lessons split up into two phases: what to do if we came down at sea, and what to do if we came down over land. The most probable situation in this second case was that we might have to parachute into Africa, for our predicted orbits took us over a great area of this vast continent. It was a slim possibility, really. But we did not want to leave anything to chance.

We did not go all the way to Africa for our homework on this subject, but we did spend a few memorable days in the sun-parched desert near Fallon, Nevada. A crew of experts in desert survival came out from Stead Air Force Base to put us through our paces. They showed us how to make protective clothing out of our parachute cloth, which looked more or less like the burnooses or headdresses worn by Arab tribesmen in the Middle East. The air was very dry, and it held fairly steady during the day at 110 degrees on a stretch of sand that often got as hot as 145 degrees. We built some crude shelters out of parachute material to protect ourselves from the fierce glow of the sun. And we learned how to conserve our energy and our body fluids in the tremendous heat. Just to see what the effects of dehydration would be, some of us intentionally went without water for several hours. We found out. Within 24 hours, our strength had ebbed to the point where it took a supreme effort just to lift a hand to wiggle a signal mirror. Even after we started drinking water again, it was hours before we regained our strength. As a result of this experience, we asked the engineers to provide us with a larger supply of drinking water in the spacecraft. There was already a tank hidden away in one corner which provided fluid for the air conditioner. It was too hard to reach, however, if we were completely worn out. So the capsule people fixed us up with one more small piece of equipment. This was a piece of thin plastic hose through which we could siphon the coolant water if we needed it.

The water we were most concerned about, however, was the sea where we would land. Let us suppose that the Mercury capsule has just returned from orbit, and now it is resting on the water. It is filled with strange sounds. The air rushes out of the cushion above the heatshield with a loud whoosh. The waves are slapping against the walls. The motion of the sea makes the capsule bob around unpredictably. You don't know for sure whether you will go on floating or not. You do know that there is a lot of help on the way. Navy ships, Air Force and Navy search planes and Marine helicopters are out tracing down the signals you have sent out. At the moment that your retro-rockets fired, about three thousand miles away, the computers went to work and pinpointed the area where you would plunk down. If you came down in an area that we had planned for, the helicopters would probably be on top of you in a matter of minutes. But in case something went wrong with the flight and you had to come down on a section of ocean where the planes would have to do some flying in order to reach you, you might have to leave the capsule, take the inflatable life raft and survival kit out with you, and wait it out.

Let us suppose now that you really are lost to the world for a short time. You have yellow dye in the survival kit to spread in the water as a signal. You also have a small mirror for flashing sunlight back at any planes you might see in the distance. If it still takes time before they find you, you can do what Navy and Air Force pilots have done for years—survive on your own. There is a solar distiller in the kit which will help you convert salt water into the pint of water you need each day to stay alive. You can rip up your suit and use the material to make some protective clothing for your face and head. You can *try* catching fish or curious seagulls. And you ought to keep yourself mentally occupied so you won't panic and do something stupid. I think I'd probably start working out algebraic problems in my head. *That* ought to keep my mind off my troubles. If you happened to get waterlogged while you were settling down, and you had not gotten *into* your raft yet, you would have another kind of problem, of course. And all the training you'd had in the centrifuge and on MASTIF would not help you here. It would be a simple case of survival. So we went through this drill several times to make sure that all of our hands understood it, and on a couple of occasions we had a minor adventure or two.

Once, when we were practicing escape procedures off Pensacola in the Gulf of Mexico, some of the swells got up to eight feet or so, and the sea was so rough that the crew of the Navy barge we were using seriously considered calling off our practice for the day. John Glenn and Scott Carpenter had a lively time of it. They climbed onto their one-man rafts and drifted clear of the Navy boats to practice the SOP we had worked out for getting out of the capsule. The waves got so high at one point that the Pensacola control tower sent a helicopter out to rescue them. John and Scott waved it away and decided to ride out the sea until the wind could blow them onto bleak little Santa Rosa Island, a few hundred yards away. The surf was so strong, however, that it upset John's raft and he went head over heels into the water. Scott caught a lucky wave just right and went skimming all the way to the beach like a surfboard rider at Waikiki. Deke Slayton took on a lot of water inside his suit when he tried to make a practice egress from the capsule with his helmet off. Because of the extra weight, he had a hard time getting into the life raft. Some swimmers that we had stationed in the water for safety helped him out, however, and he made it in good shape. We designed a special waterproof neck seal after this incident to prevent problems like this from recurring again.

These were not really close calls. We never got so far apart that we could not help out a man who was in trouble. But they did prove one thing. It takes a good deal of training to learn anything, no matter how simple it may seem. And it was just as well that we learned some of these lessons about survival before we made our flights, when we might have taken a chance on getting dunked out there all by ourselves. We learned another lesson or two on Gus Grissom's flight, and, as we will point out later on, we took some extra precautions as a result of this experience before John Glenn and Scott Carpenter went on their flights. As it turned out, John did not have to put these lessons to practical use. But Scott did, and I am sure he was glad that we had worked out the procedures ahead of time before he found himself all alone in the ocean.

GLITCHES
SAVE

John H. Glenn, Jr.

IN TIME
TROUBLE

Anyone who owns a car, a TV set or an electric toaster knows how even the simplest gadget can go radically wrong and start misbehaving. An automatic toaster, for example, will burn the bread one minute, then pop it up the next as fast as you put it in. A TV set will start sputtering one evening, work fine for an hour or so, then act up again just as you are settling down to your favorite program.

Pilots are perhaps more familiar with frustrating mechanical failures than the members of most professions. In the days of World War II, when combat planes were coming off the assembly line almost as fast as automobiles and there was a greater mathematical chance of malfunctions than there would normally be in peacetime flying, the flying fraternity blamed many of its technical problems on mythical little creatures called "gremlins." When a fuel line clogged or the control surfaces on an airplane got stuck for no apparent reason, the pilots and the mechanics alike would grumble that gremlins had gotten into the system and fouled it up. It was not always the gremlins' fault. Sometimes the blame rested on the mechanic, who had failed to check a part, or on the pilot himself, who had not taken enough time to look over his airplane before he took off. In a case like this, the gremlin was simply a convenient scapegoat. When the crews had done everything humanly possible to check their planes, however, and things still went wrong, the gremlins deserved some of the blame. Putting it on these little fictitious creatures was a handy way of admitting that complex machinery can often get the best of man no matter how hard he tries to master it.

We had similar problems in Project Mercury. In any development program which involves new and tremendously intricate hardware such as we have been working with, troubles are bound to crop up. In fact, the primary purpose of a step-by-step research like Mercury is to subject all of the systems to such a thorough evaluation as you reach each step that you are able to pinpoint the problem areas as you go, and resolve them before you get too far down the road. As test pilots, all seven of us were familiar with this kind of procedure. One reason we were brought into the program in the first place was to give the equipment the same kind of appraisal, from the ground up, that we would apply to a new aircraft we were about to fly for the first time. The Mercury equipment was unusually intricate, so we ran into some extremely complicated problems. And because we were all anxious to keep the program moving as expeditiously as possible and to reach the stage where we could actually start flying in space, some of the difficulties we encountered were definitely frustrating. Each of the problems that we had to solve meant a certain delay. These delays, in turn, seemed to raise some doubts in the public mind about the validity of the program. Most of this impatience, I think, was demonstrated by people who did not fully understand the principles of modern research and development and did not seem to realize that each failure was not necessarily a disaster, but that on the contrary every setback taught us a valuable lesson and in most cases actually improved our chances of success later on.

We did not blame any of our problems on such things as gremlins. For one thing, these creatures belonged to another era. No matter how fouled up an airplane could get, the pilot was working in an environment that was more forgiving of error and more cooperative in the sense that the pilot was not necessarily a goner just because his airplane was. In Project Mercury, however, we were preparing for a totally different environment in which the failure of a simple component in the system *could* mean disastrous results for both the mission and the crew. We were also dealing with a much more sophisticated and unforgiving machine, and our approach to it had to be relentless and sophisticated, too. So far as we were concerned, there was nothing particularly mysterious about most of our problems. They were due either to human error and fallibility, which had to be pinpointed and corrected fast, or to a technical complication which had to be ironed out before we could take another step forward.

We had our share of human errors. Despite a tremendous amount of dedication on the part of the crews and technicians who handled the complex pieces of equipment, someone could get tired or careless and make a mistake now and then. This was probably inevitable in any program that was so new and untried. An extremely complicated piece of hardware would turn up in which a couple of wires had been reversed, for example. Or someone would leave a switch in the wrong position during a check-out, and it would cause a little trouble until it was spotted. We blamed human errors like this on what aviation engineers call "Murphy's Law." Murphy was a fictitious character who appeared in a series of educational cartoons put out by the U.S. Navy to stress aviation safety among its maintenance crews. In the cartoons, Murphy was a careless, all-thumbs mechanic who was prone to make such mistakes as putting a propeller on backwards or forgetting to tighten a bolt. He finally became such an institution that someone thought up a principle of human error called Murphy's Law. It went like this: "Any part than *can* be installed wrong *will* be installed wrong at some point by someone." The law applied on a few occasions in Project Mercury, but on the whole we were spared such problems. The fact that we started from scratch in the program, trained most of the crews as we went along and made the progress we did without a serious accident of any kind attests to the skill and care of the thousands of men who made the parts, soldered the seven miles of wire together, and manned the control stations when we started flying.

Another engineering term we used to describe some of our problems was "glitch." Literally speaking, a glitch is a spike or change in voltage in an electrical circuit which takes place when the circuit suddenly has a new load put on it. You have probably noticed a dimming of the lights in your home when you turn a switch or start the dryer or the television set. Normally, these changes in voltage are protected by fuses. A glitch, however, is such a minute change in voltage that no fuse could protect against it. Glitches are a special problem when you are working with highly sensitive electronic equipment such as the many relays we had to install in the capsule to help link the complex systems together. The capsule, with its labyrinth of wiring and electronic equipment, was extremely glitch-prone and was an electronic engineer's nightmare. The crisscrossing circuits which knit all the systems into a working unit had to be plotted with great care so that a stray glitch that might arise in one circuit could not possibly find its way into other circuits and foul up another system. The eletronic impulse, for example, which explodes the posigrade rockets to separate the capsule from the booster when it is time for insertion into orbit must not be allowed to reach the separate circuit which fires the retrograde rockets that slow the capsule down immediately and start it back towards earth. If this happened, the flight could end in disaster before it ever got under way. The engineers worried about all these possibilities ahead of time, even the more remote ones, and made sure that the wiring was designed in such a way that no glitch could cross over and start a chain reaction of trouble. Needless to say, even after the capsule had been checked out hundreds of times, we still spent hours lying inside it right up to flight time, just testing circuits to make sure that there was not a glitch in sight.

The most dramatic electronic failure of the program occurred at Cape Canaveral on the morning of November 21, 1960. The flight was scheduled on the board as MR-1—for Mercury Redstone 1—and it was to be the first full-scale test of a Mercury capsule on a normal ballistic flight. The capsule was empty except for instruments. It was mounted on the Redstone, however, just as it would be for the first manned flights by Alan Shepard and Gus Grissom—and before that, by the chimpanzee, Ham. The timing of these later missions depended on the success of MR-1.

I happened to be watching from inside the blockhouse that morning as the Redstone sat puffing out liquid oxygen a few hundred feet away, so I had a grandstand seat for one of the weirdest performances Cape Canaveral has ever seen. As the countdown reached "zero" and we waited for the booster to start lifting the capsule off the pad, there was a cloud of smoke and flame, and the Redstone actually budged a little. Then its engine stopped. The Redstone settled back on the pad and the capsule, which was sitting 60 feet in the air on top of it, began to act as if it were literally blowing its top. In a fraction of a second, the escape tower mounted on top of the capsule took off in a rush of flame, a parachute came shooting out of the canister where the tower had stood, and another parachute came popping out in rapid succession and plopped down over the side of the capsule like a limp rag.

When the smoke finally cleared, the pad was littered with all kinds of debris. The Redstone still sat there, however, puffing out liquid oxygen as if nothing had happened. The capsule was still perched on top of it, trailing a parachute all the way to the ground and flashing a light on and off. Seeing all of this happen was like watching the fizzle of some gigantic Roman candle at a Fourth of July celebration.

It was late in the day before the technicians could get close enough to the pad to figure out what had happened. The Redstone was still full of fuel and liquid oxygen and its batteries were still full of power, so there was some danger that it might blow up on the pad if anyone tried to tamper with it. After waiting for the fuel to be drained out and for the batteries to run down, the crews finally got a good look at the bird and discovered what had happened. There were two plugs mounted at the base of the Redstone. One was a power plug which served to ground the booster's batteries until launch time. The other was a control plug which fed wires into the Redstone and allowed the technicians in the blockhouse to control its functions until just before lift-off. Both plugs were supposed to disconnect at the same time; they would be jerked out automatically as the Redstone started to pull away from the ground.

Both plugs did pull loose, for the booster managed to rise about an inch off the pad. But they did not pull out at the same instant. As it turned out, there were two reasons for this. One was that the prongs on the power plug happened to be one-eighth of an inch shorter than the prongs on the control plug. The other reason was that the Redstone carried a heavier load on this flight than it had in the past and therefore rose a little slower than it usually does. Therefore, instead of pulling both plugs free at the same instant, as it would normally have done, the Redstone disengaged them one at a time. The shorter one, the power plug, disconnected first. The lag between the two was very short—only about 20 milliseconds. But this was just enough to alert the electronic brain in the booster that something had gone wrong. Nothing like this had ever happened before. The result was that without the power plug to complete the normal circuit, an entirely new circuit was set up in the wiring. This was a classic example of a glitch, and the abort-sensing mechanism in the Redstone immediately shut off the engine. Under the circumstances, this was precisely what it should have done. The brain was there to sense trouble in the system and to stop the booster engine as soon as it spotted one.

From this point on, the capsule went to work and behaved exactly as *it* was supposed to. Its own electronic brain received the message that the Redstone's engines had shut down. This was exactly the same signal that the brain was prepared to receive when the booster had reached the altitude of 35 miles and it was time to jettison the escape tower and proceed with a normal mission. It was not the brain's fault that it behaved as it had been taught to do. It had not been wired, however, so that it could distinguish between two identical signals, one of which was caused by a small design discrepancy that had never cropped up before and was due, actually, to a change in operational procedures. There was only one conclusion for the capsule—the flight was over. So it fired off the escape tower. Then it began to tick off the entire sequence of "return-from-space" events, just as it was supposed to. Barometric pressure gauges inside the capsule sensed that the spacecraft was at a low altitude—sitting there on the pad, it certainly was—and automatically broke out the parachutes which normally come out at this time to stabilize the capsule and lower it gently to earth. The light beacon which was mounted on the capsule to attract the attention of recovery forces as soon as it hit the water started to flash on and off. Then the capsule began to transmit radio signals, just as it would at sea, to give the recovery fleet a precise fix on its location. The radio worked so well that two Navy P2V aircraft which were on recovery patrol duty 120 miles from Cape Canaveral picked up the signals, obediently turned around in mid-air and started to home in on the launching pad. Despite the comic aspects of the situation, this was nice going.

Some humorists at the Cape referred to this first test of MR-1 as "the day we launched the escape tower." Actually, it was a more noteworthy day than that. Once we realized that the capsule had made the best of a confusing situation and had gone on to perform its duties just as it would have on a normal flight, we were rather proud of it. It had omitted only one step, as a matter of fact, and it had a good excuse for that. Normally, the capsule would have separated itself from the booster before it started its descent and started popping out parachutes. This maneuver was programed, however, so that it would be triggered only if the capsule were pulling .05 Gs at the time of booster-spacecraft separation. Since it was subjected to a normal 1 G as it sat there quietly on the pad, the capsule had enough sense to obey its previous instructions and stand pat. If it had not done so, it would undoubtedly have taken off with the escape tower. This would have meant a very bumpy ride for the capsule and a rough landing and it might also have made it more difficult for us to determine exactly what had happened later on.

I do not mean to give the impression that we were ecstatically happy over the nonflight of MR-1. It meant another delay in a program that had already had its share of troubles. We were pleased, however, that the capsule had behaved so well under the circumstances, and we were relieved that we had uncovered this particular problem on an unmanned flight, where there was less chance of anyone getting hurt.

The Redstone people corrected the cause of the failure, and two months later we tried MR-1 all over again with the same capsule. Once more, I was in the blockhouse to watch the launching, and this time it went fine, without a hitch. Looking out through the thick, heavily reinforced windows, I saw the Redstone begin to rise, slowly and unwavering in a burst of flame. The blockhouse trembled slightly from the vibration as the booster gathered speed. When it got to an altitude of about 700 or 800 feet it seemed directly over our heads, and all I could see were the points of fire stabbing out of the tail and back towards earth. Then it passed out of sight, and I stepped back from the window to listen to a voice on the intercom circuit which gave us a step-by-step progress of the flight. The booster and capsule reached a speed of more than 4,000 miles an hour and an altitude of about 35 miles before the booster engine cut off. Then the escape tower jettisoned on schedule. The capsule and the Redstone separated from each other at this point, and the suborbital flight began. The capsule soared weightless for six minutes as it reached the top of its great parabolic arc. Then, at just the right moment, and at an altitude of about 135 miles, the retro-rockets fired. The spacecraft began its descent, and when it reached an altitude of 21,000 feet, a stabilizing parachute—the drogue chute—popped out automatically. At 10,000 feet, the huge orange and white main chute blossomed out. This time the parachutes did not pop out in vain. They filled with air and did their job. A search plane working for the recovery fleet spotted the capsule as it floated to earth and watched it splash into the ocean, 230 miles off the Florida coast. Helicopters flew to the spot and one of them hooked on and lifted the capsule out of the water. Forty-eight minutes after the launching at Cape Canaveral, the capsule was safe on the deck of the carrier *Valley Forge*.

We were almost ready now for Ham, who would be the passenger on MR-2, the next major test in the series. First, however, we had to make a slight correction in the flight plan. Even though the second MR-1 had performed according to schedule, we had had a few moments of concern about it during the flight. While the Redstone was still climbing and heading towards the east, it ran into some high-altitude winds which were blowing from the east and which slowed it down just a bit on its trajectory. It went on to hit its target, but it tarried just long enough over our heads in the Cape area to stretch the local safety factor almost to the limit. We always aim, on any launch, to get the missile downrange as soon as possible so it will not endanger a populated area if it has to be destroyed. In addition to this, we also realized that by letting the Redstone climb so steeply on the first leg of its flight, we were letting the passenger in for a very steep descent on the other end. It would be like zooming down from the highest point on a rollercoaster, and it would mean pulling 12 Gs or more during re-entry. This was not only a little grueling but it was also unnecessary, so we lowered the gun barrel for MR-2 and aimed for a flatter trajectory. On this flight, the capsule would rise to an altitude of only 115 miles—some 20 miles shorter than MR-1's apogee—and would make up for the difference by landing 290 miles downrange instead of 230. This meant shifting the recovery fleet further out into the ocean and inserting the changes into the computers and electronic programers which would control the mission.

We were well aware along about this time that the country in general was getting slightly restless over the delays. Politicians were making statements to the effect that we were not moving as fast as they thought we should. Journalists kept pointing out that the Russians were running fast and winning the race. One critic suggested sarcastically that if Lindbergh had planned his flight across the Atlantic the same way we were planning our space flights, he would have sent the "Spirit of St. Louis" across empty first and then with a monkey in the cockpit before *he* ever tried it. This was a clever remark, but it was beside the point. The role of man in space at this juncture was to test his ability to cope with this new environment and to maneuver the machine which put him there. There was no point trying to do that if the machine was not ready for the man. Lindbergh did an extremely daring and imaginative thing. He flew alone in a single-engine airplane across a vast expanse of water that had never been charted for this kind of crossing. At this particular juncture in the development of flying, it was an epic achievement and a tremendous breakthrough in man's ability to conquer his environment. Our problems were of a rather different kind, however. Lindbergh did have wings to rely on. And the airplane itself had proven that it could fly. In our case, the first Astronaut was going to sit for the first time on top of 76,000 pounds of volatile rocket thrust and take off in a wingless craft that no one had ever flown at all. He was going to submit himself to a new environment and to various physical stresses that no American had ever experienced. The most important difference was that we were engaged in a national scientific effort in which any short cut that we took meant that we would have to leave out a step or two and learn that much less. We were not in this program to perform great space feats, but to gather valuable information for the future.

We did get a little restless ourselves, however, from time to time. We all knew what an early success would mean in terms of national prestige and confidence. And I suppose that some of the criticism did get under our skin. Project Mercury had been laid out, however, as an orderly series of tests, each one of which taught us certain lessons and provided us with certain data which would be fed into the next test. If we tried to interrupt or short-cut this process by jumping one of the important steps in the development program, we might have achieved what seemed like a success or two at the time. We would not have proven very much in the long run, however. It was far more important for us to do it right, we felt, than to do it first.

I am sure that there were quite a few aspects of the program that it was difficult for outsiders to understand. NASA announced early in the program, for example, that it hoped to launch an Astronaut on a ballistic flight some time in 1960 and put one of us into orbit by some time in 1961. NASA added, of course, that this schedule depended on the successful completion of all of the preliminary tests. In one five-month period in 1960, which led right up into late December, three out of four of these tests were failures. One of these was the first MR-1, which blew its top. Another was the first flight of a Mercury Atlas—MA-1—which was supposed to hurl an empty capsule 1,400 miles downrange to test its ability to withstand the maximum heat of re-entry. The Atlas took off from the pad in good shape, but one minute later it blew itself up, and both the capsule and the booster fell back into the ocean in bits and pieces. The engineers decided, from studying telemetry and photographic evidence, that the failure was caused by excessive vibration. There were two reasons for the vibration. One, as we have explained earlier, was that the capsule had to be constructed in an unusual shape from the standpoint of streamlining and aerodynamics so it would slow itself down as it came back through the atmosphere and not burn up. This also meant that the capsule would not have a particularly smooth ride on the way up, especially with the escape tower stuck on top of it. In addition, the fairing or adapter section which mated the capsule to the booster added somewhat to the roughness of the flight. The engineers altered the adapter, strengthened the skin of the Atlas so it could withstand the extra vibration and tried again. Seven months after the failure of MA-1, MA-2 was launched and was a success. The fixes the engineers made did the job, and we never had the same trouble again. It was a good example of the trial-and-error nature of a new program, and of the ability of the engineers to learn from a single failure and prevent it from occurring again. The third of the three failures occurred at Wallops Island when a Little Joe was launched with the first instrumented capsule aboard. Because there was no ASIS on board this particular flight and the capsule and rocket were not attached together with standard equipment, they failed to separate and plunged together into the Atlantic from an altitude of 53,000 feet and sank beneath the waves just 13 miles from where they had started.

Each one of these failures meant a delay in the program while the engineers determined the reasons for it and in some cases went back to the drawing boards to work out a solution. Sometimes, if we were lucky, the job could be done in a matter of weeks. More often it took months. There were no short cuts, and the engineers would not have taken them if there were. You do not build a house any faster by leaving out chunks of the foundation to save time—not, that is, if you plan to live in the house for years to come.

I suppose the most frustrating thing about the delays, for the public at least, was the fact that our officials had to announce schedules which we did not seem able to keep. There was no easy way out of this problem. Setting up schedules for such a complex project is a risky business. The people who announce them have only two choices. They can greatly exaggerate the time they believe it will take them to get the job done—and look good whenever they manage to beat the deadline—or they can announce a series of realistic target dates which they really hope can be met if everything goes reasonably well. This is the most honest technique, and it is the one that NASA followed. The chief virtue of this approach is that it gets everyone in the program geared up for an early deadline and doing their best to meet it. The chief drawback is that when you do run into trouble the delays are more apparent. Their obviousness does not mean, however, that the project itself is failing.

As Robert Gilruth, the director of Project Mercury, pointed out during one long period of troubles and delays, we could have achieved our goal much sooner if we had been interested only in putting a man up on a Redstone. If a simple ballistic flight had been our only goal we could easily have jumped a few steps and accomplished the feat long before we did. But the entire system was designed for orbital flight, for the long missions which would get us deep into space and give us time for lengthy observations. The ballistic flights were necessary steps in this direction. They checked out the capsule and its systems, and they provided training for the pilots and the ground crews. There was no point in rushing the Redstone flights, however. In order to learn the greatest number of lessons even from these early missions we wanted each capsule we launched to be as nearly like the capsule we would use in orbit as possible. To do otherwise would have been a foolish waste of time and money. This is one reason we took so long. We wanted every flight to count, and we wanted to be right when we were ready. Someone estimated that we spent a good two-thirds of our time devising and testing out the system of automatic controls which would allow us to orbit a capsule without a man before we were ready to risk a pilot. It so happens that safety was always one of the most important factors in running Project Mercury. There were many unknowns involved in orbital flight, and we needed a system that could fly the capsule automatically until we could get a toe-hold in space and gain enough experience to be able to dispense with many of the automatic devices. We will not use these extra systems nearly so much on future flights as we have during the test program. The automatic systems had to be checked out thoroughly, too, of course; and we decided to do this on the ballistic missions even though there was no requirement for automatic controls on these flights. One of the real values of the U.S. approach to space is that it involves a broad-based program that proceeds from one step to another and builds on it the way a mason builds a brick wall. We have pulled no stunts just for the sake of stunts nor have we launched a single flight that did not take into account lessons we had learned on previous flights and provide previews of things we intended to try in greater detail on future missions. All of the various events tie together and lend strength to each other. The X-15 program, for example, which has been run by NASA concurrently with Project Mercury, has the object of testing out an entirely different concept of flight under controlled conditions that utilizes the lifting principle of a wing. Eventually, we hope to combine the best features of the X-15 idea with the other lessons we are learning about space flight in Project Mercury and form a different kind of vehicle for future space exploration. In the meantime, we are also learning other things from balloonists like Dave Simons who have studied the composition of the upper atmosphere and the edge of space beyond it. All of these studies go hand in hand. Some statistics that came out shortly after my orbital flight indicate that this across-the-board approach has paid off extremely well. At the time the statistics were compiled, the United States had launched a total of 68 satellites, had recovered 15 of them and had 34 of them still in orbit, of which 10 were still transmitting data. As of the same date, the Soviet Union had launched 13 satellites, had recovered five and had one still in orbit which was no longer transmitting.

All this was of little encouragement early in the program, of course, when we seemed beset sometimes by failure and delays. It was not an easy time for any of us. We all felt that we could use some forty-eight-hour days and ten-day weeks to get our work done. As our director, Robert Gilruth, pointed out on one occasion when critics of the program were asking for more and faster progress, we fully expected to wake up some morning and find that the Russians had beat us to it. We estimated that they had a head start of at least four years over the United States in space preparations. We did not see how we could overcome that kind of advantage overnight, and of course we did not. The Russians did beat us to it as far as getting up there first was concerned.

The delays did not affect our training, however. We went on with our practice runs on the centrifuge and on MASTIF and with the airplane flights that tested our reactions to weightlessness. Our sharpness only increased with time.

So, I assume, did Ham's. On the morning of January 31, 1961, Ham was launched. The scientists had decided to use a chimp as a stand-in because this particular animal closely resembles man from a physiological standpoint and can be trained to take part in a scientific experiment. A chimp's reaction to various stimuli, incidentally, is seven-tenths of a second, which is fairly close to the average man's reaction time of five-tenths of a second. Ham had been taught to perform a few simple tricks in order to test his ability to carry out useful functions during the four and a half minutes of weightlessness that he would be subjected to. He watched a series of flashing lights and pulled various levers in sequence when the lights came on. A supply of water and banana-flavored pellets was installed inside his chamber to reward him for his prowess on the levers and keep him happy. He was strapped into a small couch which resembled the contour couch we would use, and he was encased in a pressurized plastic chamber about the size of a small trunk which was fastened in place inside the capsule and connected with the oxygen supply. The chamber was sealed off from the cabin of the capsule, just as a man would be sealed off inside his pressure suit. Ham also had medical sensors attached to his body, much as we would, to record his heart rate, respiration and body temperature during the flight.

Ham could not know it—or do much about it if he did—but his flight did not go exactly according to plan. To begin with, the launching was delayed for two hours while the launch crew replaced a broken spring on the hatch. An oxygen valve jiggled loose inside the capsule during the launch—apparently from the vibration of the blast-off—and this kept the interior of the capsule itself from pressurizing. Fortunately, the chamber that enclosed Ham worked fine and it kept him sealed up and in good shape during the flight.

These were fairly minor problems, however, compared to a combination of events which took place a few minutes after Ham left the launching pad. The Redstone booster had too much oomph and it built up 1,000 miles per hour of extra speed. The electronic brain sensed this abnormality and triggered the escape tower which fired its rocket and gave the capsule an additional boost of 800 miles per hour. As a result, the capsule went 130 miles further downrange than it was supposed to go and landed 420 miles from the launching pad instead of the planned 290. It took quite a bit of churning on the part of the Navy's propellers to steam towards Ham for the recovery, and it was two hours before the nearest ship reached him. The crew discovered when the capsule was lifted out of the water that the heatshield on the bottom of it had broken loose on impact. It had already done its job of fending off the heat of re-entry, so no harm was done on that score. But the capsule bobbed in the water at an angle, allowing sea water to leak in through a pressure regulator in the capsule wall. When the hatch was opened, water was already sloshing around at the base of Ham's chamber. It was watertight, however, and Ham did not get wet. As a result of this flight, the engineers learned a few more valuable lessons and made some changes in the life-support system so we would not have the same problem with pressurization on future flights. One valve was redesigned so it would not stick, and an extra valve was installed in the oxygen system as a safety measure to hold the cabin pressure at a more even level.

I go into all these problem only to point out that our later successes were due in great part to the fact that we spotted the glitches and other difficulties when we did and ironed them out. Each failure led to a remedy which made the system better and safer for future flights. Every bug that cropped up in the preliminary tests was one less problem we would have to face later.

At the same time that we were experiencing some of our more spectacular and discouraging failures, we were also having our share of successes. Sometimes the bad days seemed to outnumber the good days, but in the long run the patience and determination of the managers of our program—men like Bob Gilruth and Operations Director Walt Williams —paid off. We knocked over quite a few hurdles as we prepared ourselves and learned the new techniques. We set the hurdles up and knocked over a few again. We knocked over fewer hurdles with each new test, however. We were learning how to do it, and considering the timetable we had set ourselves, we were not doing too badly at all. Project Mercury began in 1958. The first successful flight of a manned Mercury spacecraft came in 1961. It normally takes from three to five years to design a new airplane and get it off the ground—even though we have been in the air age for some time and know a lot more about the problems involved. In the case of Mercury, we worked out an entirely new kind of flight system under vastly different conditions. We did not know much about the system at all when we started to work on it, and yet we had it off the ground in less time than it usually takes to develop a new plane. You cannot argue with that kind of success.

During the delays, one critic of the program went so far as to suggest with obvious sarcasm that perhaps the best way for us to guard against failure in our man-in-space program was not to try to put a man up at all. We had news for him, however. Al Shepard was getting ready to go.

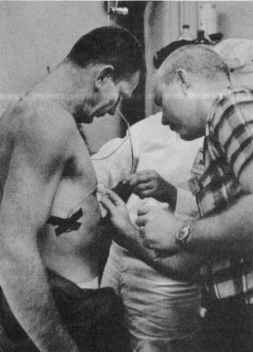

There was much to learn, and for the first year the Astronauts buckled down and acquired a basic education in subjects like astronomy, physics, astronautics and physiology. Like Deke Slayton [ABOVE, RIGHT], who is having biological sensors taped to his body by Dr. Douglas, they went through a series of tests to learn how much stress and strain their bodies could stand. Like John Glenn [RIGHT], they holed up in their rooms and pored through stacks of technical literature. But while they were preparing themselves for their missions, they were also paying close attention to the development of the hardware that would put them into space. This involved long months of trial-and-error research, and occasionally—as happens with any new development—unforeseen setbacks. One of the most spectacular fizzles occurred on November 21, 1960, when a Redstone missile with an empty Mercury capsule mounted on top of it blew its stack [ABOVE] and put on a display of pyrotechnics that confounded the engineers. Two months later, the same experiment was tried again, and this time the Redstone and capsule took off in perfect shape.

THE FIRST

Alan B. Shepard, Jr.

AMERICAN

I HONESTLY never felt that I would be the first man to ride the Mercury capsule. I knew I had done well in the tests and the simulator rides. And I thought there was a good chance that I would get one of the early flights. But I had conducted my own private poll and frankly I figured that one of the others would probably go first. I got the word that I had been selected in the combination office and classroom at Langley where we had listened to so many of our lectures and had done a lot of our homework. The only people present for the occasion were the seven of us Astronauts and Mr. Robert Gilruth, the director of the Space Task Group.

There was no hemming and hawing around. Mr. Gilruth simply announced the choice of pilot and backup pilot. After my name was read off I did not say anything for about 20 seconds or so. I just looked at the floor. When I looked up, everyone in the room was staring at me. I was excited and happy, of course; but it was not a moment to crow. Each of the other fellows had very much wanted to be first himself. And now, after almost two years of hard work and training, that chance was gone. I thanked Mr. Gilruth for his confidence. Then the others, with grins on their faces covering up what must have been their own great disappointment, came over and congratulated me.

That night I told my wife, Louise. I informed the rest of my family in plenty of time for them to get set for the press and the publicity. They all kept the secret very well. It had to be a secret in case something went wrong and we switched signals.

There was nothing secret about all of the work we had to do, however. The public has heard a lot about countdowns, and I would not blame people if they got the idea that the pressure does not start until someone starts counting from ten down to zero. Actually, everything is a countdown. You are always aiming at some specific launch date on the calendar. And the countdown really begins when they put the first two pieces of metal together back at Detroit or out in St. Louis months before the launch is even scheduled.

For me, the countdown started with quite a bit of homework. I looked over the Mercury Redstone missile Number 7 which would propel me off the pad and talked to the technicians who had put it together. I made a study of the abort-sensing devices which might have to spring the spacecraft free of the booster if it acted up. And I stuck close to the capsule as the workmen put it together, piece by piece, and then hung around to watch as the engineers tested it, section by section. I was more familiar with that capsule when it was all over than I have ever been with any other piece of hardware in my life.

A capsule is quite a bit like an automobile. No two identical models ever drive exactly the same. Each one has its own delicate differences and small deviations. Sometimes an ammeter in one capsule will read zero, when actually it is pulling two amps. In another capsule it may be just the reverse. You can cope with these deviations if you know about them ahead of time. It's like driving a car with a soft brake or a little play in the steering wheel. You have to allow for it. There was nothing seriously wrong with Number 7. I did detect a slight difference in the way the control stick responded when I moved it in some test runs. The stick seemed to be a little stiffer than the ones I had gotten used to in the trainers. But there was nothing alarming in this. Actually, the capsule checked out very well. The technicians knew that Number 7 would be manned, and they gave it a lot of special attention.

To make sure that all of the equipment was in working order and that all of the personnel involved were completely familiar with the techniques they would use to get us off the pad, into space and back again, I went through 40 separate simulated flight tests or dry runs in which everyone rehearsed his own role, from beginning to end, and kept an eye on all of the gauges, signal lights and fuses which controlled each sequence and showed how the gear was working. In addition to checking out the equipment and the crews, I was also checking myself out—and so were the NASA doctors. As the first American to go through this sort of thing, I had to pay an unusual amount of attention to my own inner thoughts and to my psychological bearing as the pressures began to mount.

I tried to pace myself—which is difficult for me to do. The only way for me to do this was to realize when I was getting too involved in one concept, and to fall back and regroup. Once in a while the thought of making an unsuccessful flight would get to me too much and I could feel it in my stomach. I was not going around shaking, but there were some butterflies. There was so much work to do, however, that I never took time out specifically to worry.

For as long as I could, I tried to avoid moving into the hangar. Even though we were spending most of our time working on the Cape, I stayed at a motel outside the gates just to keep from gearing up all the technicians and other people inside who would have to support us as long as we were there. I knew we would need them later. And I tried to play it cool so that they would not get overanxious and make mistakes. Finally, things became impossible at the motel—strangers and reporters kept barging in on us—so I quietly moved out and settled down in Hangar S for the duration. It was very pleasant out there. We had well-appointed quarters and plenty of peace and quiet which allowed us to concentrate on our work.

Hangar S is one of the principal NASA buildings in the industrial area on Cape Canaveral and houses, among other facilities, the big, sterile "white room" where the spacecraft is fussed over, weighed, balanced and given its final inspection and tests before it goes out to the launching pad to be mated to the booster. On the second-floor mezzanine of the hanger is a suite of rooms set aside for us Astronauts. The crew quarters are in room S205. This is an air-conditioned room about 20 by 30 feet in size which is furnished with two double-deck bunks, two beige-colored nylon couches, a reclining chair, magazine racks, a desk and chair, radio, TV and a clock. The room was decorated by the nurse who had been assigned to us, Air Force Lieutenant Dolores ("Dee") O'Hara. Dee picked colors which she decided were relaxing and restful. She had the walls painted robin's-egg blue and covered the two windows from ceiling to floor with champagne-colored drapes. The next room down the corridor is for conferences. Next to that is the pressure-suit room, where we suit up for our missions. And next to that is the aeromedical room, where the doctors keep their charts, thermometers, sensors and probing instruments. On the morning of May 2, 1961, when I was first scheduled to go, a heavy rain was falling outside and flashes of lightning were playing around the launching pad three miles away. The signs were not propitious, and at 3:30 A.M., with the liquid oxygen fuel already loaded aboard the booster, the technicians took a look at the lightning and declared a hold. They started working again at 3:50, with the count at T minus 290 minutes.

I frankly did not think we would go that morning. I was not trying to second-guess anyone, but the weather did not look good at all. I was sure we would not get the results we needed even if we did go. But the crews were ahead on the countdown, and if we did not try it that morning we would have to go through a long 48-hour delay before we could refuel the Redstone and try again.

I slept until about 1 A.M., when Dr. Bill Douglas woke me and told me they had decided to put the liquid oxygen in and that it was time to get up. I had orange juice, a filet mignon wrapped in bacon and some scrambled eggs for breakfast. Gus Grissom and John Glenn joined me and had the same, except that they asked for poached eggs. Then I went through my final preflight physical. The NASA doctors had had a longer session with me on the day before and they reported that I "seemed relaxed and confident."

Lieutenant Colonel Douglas, the Air Force doctor who had cared for the Astronauts and paid special attention to each of them as the date of his mission approached, knew his seven patients as intimately as anyone else in the country. "The morning each man went into space," he has said, "was the only time when he was not normal, when he was a superman. No one else in the world could have gone to bed that night, had a sound sleep, and then gotten up bleary-eyed the next morning when I went in to wake him and acted just as if he were going out duck-hunting or starting on a fishing trip."

Douglas and his assistants had already gone over every inch and pore of Shepard's body. They noted that the nail on the fourth toe of his left foot was coming loose because someone had accidentally stepped on him. His back was peeling slightly from a recent sunburn. And there was a slight "eruption" surrounding a two-millimeter "tattoo" which had been painted on Al's chest to mark the spot where one of the sensors would be placed. Outside of that, Shepard was in the same perfect physical shape that he had been in all along. His eyes were normal. His ear canals were clear. The X-rays and the electrocardiogram were all normal. His thyroid was "smooth and symmetrical, with no tenderness." The doctors asked him to murmur "ninety-nine, ninety-nine" a few times while they listened to his chest, and then tapped his back to look for respiratory ailments. They found none. And then, when they whanged up a tuning fork to a low register of 126 cycles per second and stuck it against Shepard's forehead, Al heard nothing unusual in either ear. This was fine, because he was not supposed to.

Then a NASA psychiatrist talked with Shepard in the hangar and made his report. "He discussed the potential hazards of the flight realistically," the doctor wrote, "and expressed slight apprehension concerning them. However, he dealt with such feelings by repetitive consideration of how each possible eventuality could be managed. His thinking was almost totally directed to the flight. No disturbances in thought or intellectual functions were observed."

With all of this taken care of, Shepard—who had already slid into the bottom part of his specially ventilated long-john underwear—submitted quietly while the doctors positioned the six sensors which would help transmit his physiological functioning back to the control center during the flight. The sensors went under the right armpit, on the upper and lower chest, on the lower left side, in the rectum (for temperature) and below the nostrils (to measure respiration). The loose wires dangling from the sensors were led to a common terminal so they could later be plugged into a socket mounted near the right knee of his pressure suit.

Wearing white socks and white canvas slippers in addition to the padded underwear and the sensors, Shepard walked into the dressing room and started to put on his pressure suit. Including the helmet, it weighed about 25 pounds. Joe Schmitt, the NASA fitter, helped to steady Al while he stuck first his left foot, then his right, into the legs of the suit and threaded the sensor wires through a hole on one thigh. Then Al stood up, wriggled into the rest of his rubber and fabric skin and zipped it up across the chest and around the middle. He pulled on his boots and gloves, zipped the gloves to the sleeves of his suit, slid a pair of plastic galoshes over the boots to keep dirt off them until he reached the capsule, and held steady while Schmitt lowered the space helmet over his head and clamped it to the suit with a ring lock.

*Completely encased in his heavy cocoon, Shepard climbed awkwardly
onto a contour couch and rested while Schmitt inflated the suit with five
pounds per square inch of oxygen and proceeded to test it for leaks.
As each pound went in, Al held up his right thumb, gladiator-style, to
indicate that everything was in order. His faceplate was closed now, and
he could communicate with the other people in the room only through
the tiny microphone anchored in front of his lips. The suit checked out.
Schmitt deflated it and attached a black portable air-conditioner to keep
Shepard cool until he could reach the capsule and plug himself into the
oxygen and air-conditioning system there. Shepard held the box in his
hand, ready to shuffle off to the transfer van which was parked just outside the hangar door.*

*As Shepard suited up, the launch crews at Pad 5 were still working
away on the capsule and the Redstone. Except for the weather, there
were no delays. As soon as it was obvious that Shepard was in shape to
make the flight and that he would not be needed as a backup pilot, John
Glenn went out to the pad to help prepare the capsule. Dressed in white
coveralls and a white hat shaped like a butcher's cap, John went up the
gantry elevator to the big plastic-enclosed platform which surrounded
the capsule. The room was sealed and air-conditioned by a 10-ton machine which kept out dirt and controlled the heat. And because it reminded them sometimes of a Miami resort, the technicians who worked
there called it "Surfside 5," after a TV program laid in Miami and known
as "Surfside 6."*

*John Glenn reported that he found the capsule crew raring to go.
On the beaches a few miles away, people were huddling under ponchos
and umbrellas, determined to sweat out the launch despite the rain. And
out at sea, some 300 miles downrange, the ships of the recovery fleet reported that they were on station, ready to launch jet planes which would
follow the capsule on its descent, and helicopters which would fish Shepard and his capsule out of the water after they had landed. On board the
biggest ship in the fleet, the U.S. Navy carrier* Lake Champlain, *was an
emergency crew which normally stood by to chop pilots out of any
planes that crash-landed on her deck. Now this same crew went through
a final rehearsal of the special procedures it would use to get Shepard out
of his capsule if for some reason his hatch failed to pop open. A few of
the officers on board had served with Shepard in the Navy, and they
knew him well enough to be especially anxious as the hour drew closer
for him to land near their ship. That night, as the sea grew dark, the*

ship's chaplain came on the loudspeaker system, and a hush fell over the ship as he said a prayer which spoke for the nation.

"Gracious God," the Methodist chaplain said, "as a precious human life is propelled into the heavens, we become apprehensive, tense and fearful from anticipation of imminent danger. We are thankful there are men who in their dedication are willing to prepare our way and open the door into space by sacrificing their very lives if necessary. May success crown our efforts to explore and develop paths to achieve not only an expanding universe, but a safe and more peaceful one in which we may live with ourselves and with Thee. Amen."

At 5 A.M., just as the sky was beginning to get light out over the Atlantic, a small group of reporters and photographers who had been allowed onto the Cape under a military escort lined up outside Hangar S, waiting for the Astronaut to appear and make his way to the transfer van. They did not know which of three men he would be—Shepard, Glenn or Grissom. This had not yet been announced to the public. Television lights were on, and cameras were ready to roll. Al Shepard stood in his suit just inside the hangar door and was about to walk through it and make his debut when someone suddenly called for him to wait. This was the first real indication that the mission was being scrubbed. Al took off his gloves and sat down to relax. Walt Williams and Bob Gilruth, the operations director and director of Project Mercury, had already explained to him that the weather gave them only a 50-50 chance of going ahead with the launch. They had gambled on these odds when they started fueling the Redstone with liquid oxygen shortly after midnight. Since the fueling operation marked a point-of-no-return for the technicians—the missile would now have to go through a 48-hour regeneration period if the mission was scrubbed—the men in charge decided to carry on and hope for a lucky hole in the sky through which they could shoot. The hole never appeared, however, and at 7:25 the mission was cancelled. Shepard inched his way out of his space suit, and someone handed him an ounce-and-a-half of brandy which he drank.

"He didn't really need it," said Lieutenant Colonel John A. ("Shorty") Powers, the Astronauts' public information spokesman. "There were about nine of us there," Powers added, "who needed it more than he did. He just joined us."

Within the hour NASA decided to break the news on who the man was who had suited up and almost gone.

I was relieved when they made the announcement. It was getting to be a strain keeping the secret.

That afternoon, while the technicians drained the corrosive fuels out of the Redstone with blasts of hot air and rechecked all of the circuits for another try the next night, I took a 15-minute nap, answered some of my mail and read the congratulatory wires. Then I drove out to a deserted beach to do some running to keep in trim, and went to Surfside 5 to see the capsule.

This time, things looked very good. There was a real feeling of "go." At the scheduling meeting Thursday morning we got pretty fair weather reports. The launch crews were picking up the count again at T minus 390 minutes, and I felt glad that I was going to be able to give it a whirl. There were some serious things that had to be said at the meeting, but there were some good jokes, too. Late that evening, after I made a few simulated runs in the trainer as a last-minute refresher, John and I went out on the beach to chase a few crabs around. Things were businesslike in the hangar that night. There was no "Just Before the Battle, Mother" atmosphere at all. Everyone realized that there were strong feelings present, but they were not brought into the picture. I called Louise at our home in Virginia Beach, talked to my older daughter, Laura, who was at school in St. Louis, and to my parents in New Hampshire. We had roast beef for supper that evening, and I went to bed at ten without bothering to shower. John Glenn slept in the same room, and before we turned in we went over a few last-minute changes in the recovery-ship signals so I would know what to expect. I went off to sleep in 10 or 15 minutes. There were no dreams or nightmares or charging around on the bed. I woke up once, about midnight, and went to the window to check on the stars. I could see them, so I went back to sleep.

At a little after 1 A.M. I got up, shaved and showered and had breakfast with John Glenn and Bill Douglas. We joked about what a tough life it was having to eat filet mignon again for breakfast. John was most kind. He asked me if there was anything he could do, wished me well and went on down to the capsule to get it ready for me. The medical exam and the dressing went according to schedule. There were butterflies in my stomach again, but I did not feel that I was coming apart or that things were getting ahead of me. The adrenalin was surely pumping, but my blood pressure and pulse rate were not unusually high. A little after four, we left the hangar and got started for the pad. Gus and Bill Douglas were with me. Dee O'Hara, our nurse, went as far as the hall with us, and I said, "Well, Dee, here I go."

They appeared to be a little behind in the count when we reached the pad. Apparently the crews were taking all the time they could and being extra careful with the preparations. Gordon Cooper, who was stationed in the blockhouse that morning, came in to give me a final weather briefing and to tell me about the exact position of the ships. He said the weathermen were predicting three-foot waves and 8-to-10-knot winds in the landing area, which was within our limits. We had a device in the van to check on the sensors, and everything was working fine. I rested my weight in a reclining chair while all this was going on.

Along about this time Shepard and Grissom went through a little routine to relax everyone in the van. Al had been much amused by an "Astronaut" recording which featured a Spanish Astronaut named José Jiménez. When someone asked José in an interview what he was going to do on his epic flight, he answered in a poignant Spanish accent, "I'm going to cry a lot." (Al saved that line and used it on Gus when he went up—see the following chapter.) At one point in the brief routine, Shepard —still imitating the Jiménez record—started to tick off in his own best Spanish accent the qualities a good Astronaut must have. He listed the obvious ones, like courage, perfect vision and low blood pressure. Then he added, "And you got to have four legs."

"Why four legs?" asked Gus, who was playing the straight man.

"They really wanted to send a dog," Al answered, "but they thought that would be too cruel."

Out on the beach an anxious crowd was gathering. This time, thanks to the improvement in the weather, it was at least triple the size of the group which had assembled two days earlier. A waning half-moon slid in and out behind dark, spooky clouds. But the stars were visible, and for miles around the residents of the area knew that everything was going according to schedule. They could tell, because the otherwise dark sky was pierced by brilliant lances of bluish light which came from the batteries of big searchlights that had been drawn up around the launching pad to illuminate the work. The people on the beaches were the "bird-watchers" of Cape Canaveral. Most of them had seen many "birds," or missiles, go. As the hours wore on, they were patient and optimistic. One experienced lady, who had brought along blankets, a jug of coffee, a transistor radio and a pair of high-powered binoculars, was both excited and unworried.

"There'll come a time," she said, gazing out at the searchlights in the distance, "when space travel will be as common as jet planes are today."

Shortly after five, and some two hours before lift-off was scheduled, Al asked to leave the van and ride the gantry elevator up to Surfside 5. He said he wanted some extra time to visit with the launch crews and to check over the capsule. With Grissom standing next to him, Shepard paused at the gantry base to take a good look at the Redstone before he climbed on top of it.

I sort of wanted to kick the tires—the way you do with a new car or an airplane. I realized that I would probably never see that missile again. I really enjoy looking at a bird that is getting ready to go. It's a lovely sight. The Redstone with the Mercury capsule and escape tower on top of it is a particularly good-looking combination, long and slender. And this one had a decided air of expectancy about it. It stood there full of lox, venting white clouds and rolling frost down the side. In the glow of the searchlight it was really beautiful.

After admiring the bird, I went up the elevator and walked across the narrow platform to the capsule. On the way up, Bill Douglas suddenly handed me a box of crayons. They came from Sam Beddingfield, he said. Sam is a NASA engineer who had developed a real knack for relaxing all of us, and I appreciated the joke. It had to do with another fictional Astronaut who discovered just before he was about to be launched on a long and harrowing mission that he had brought along his coloring book to kill time but had forgotten his crayons. The guy refused to get into the capsule until someone went back to the hangar and got him some. I laughed and handed the crayons back to Douglas. I said we would be a little too busy on this trip to use them. When we got to the top, the preparations were almost completed.

Gus Grissom later described to a friend a plot which he had had in mind that morning but never carried out.

"Test pilots," Gus explained, "have a gruff saying which they like to use sometimes when they see a buddy going out to wring out a new airplane. 'Go blow up,' you tell him. It sounds cruel, but not to the other test pilot. He knows he may blow up anyway, and you're just using it as a little joke to help him relax. It usually works. On the morning when

Al Shepard was going, and I was with him all the time, I had it in the back of my mind to say this to Al—just as he went up the elevator to crawl into the capsule. I knew he would have laughed, and it might have helped chase away the butterflies. But when the time came, I had a few butterflies myself. I couldn't say it."

I walked around a bit, talking briefly with Gus again and with John Glenn. I especially wanted to thank John for all the hard work he had done as my backup. Some of the crew looked a little tense up there, but none of the Astronauts showed it.

At 5:20 I disconnected the hose which led to my portable air-conditioner, slipped off the protective galoshes that had covered my boots and squeezed through the hatch. I linked the suit up with the capsule oxygen system, checked the straps which held me tight in the couch, and removed the safety pins which kept some of the switches from being pushed or pulled inadvertently. I passed these outside.

John had left a little note on the instrument panel, where no one else could see it but me. It read, "No Handball Playing in This Area." I was going to leave it there, but when John saw me laugh behind the visor he grinned and reached in to retrieve it. I guess he remembered that the capsule cameras might pick up that message, and he lost his nerve. No one could speak to me face to face now. I had closed the visor and was hooked up with the intercom system. Several people stuck their heads in to take a last-minute look around, and hands kept reaching in to make little adjustments. Then, at 6:10, the hatch went on and I was alone. I watched as the latches turned to make sure they were tight.

This was the big moment, and I had thought about it a lot. The butterflies were pretty strong now. "O.K., Buster," I said to myself, "you volunteered for this thing. Now it's up to you to do it." There was no question in my mind now that we were going—unless some serious malfunction occurred. I had anticipated the nervousness I felt, and I had made plans to counteract it by plunging into my pilot preparations. There were plenty of things to do to keep me busy, and the tension slacked off immediately. I went through all the check-off lists, checked the radio systems and the gyro switches.

At other places around the Cape at this point, the other Astronauts were taking up their positions to back me up in any way they could. Deke Slayton sat at the Capsule Communicator (Cap Com) desk in Mercury Control Center. He would do most of the talking with me during the countdown and flight so that both the lingo and the spirit behind it would be clear and familiar. John Glenn and Grissom joined Slayton as soon as I was firmly locked in and there was nothing more they could do at the pad. Wally Schirra and Scott Carpenter stood by at Patrick Air Force Base in their flight suits, ready to take off in two F-106 jets to chase the Redstone and capsule as far as they could and observe the flight. Gordon Cooper stayed in the blockhouse, monitoring the weather and standing by to help put into effect the rescue operation he had worked on which would get me out of the capsule in a hurry if we had an emergency while we were still on the pad. On the big map inside Mercury Control, which showed the locations and status of the tracking stations, recovery ships and communications net, all the lights were green. All conditions were "Go." The gantry rolled back at 6:34, and I lay on my back 70 feet above the ground checking the straps and switches and waiting for the countdown to proceed.

I passed some of the time looking through the periscope. The view was fascinating. I could see clouds up above and people far beneath me on the ground. When the sun rose, it came right into the scope and I had to crank in some filters to cut down on the glare.

Shepard had a long, long time to admire the view. For the launch crews ran into some trouble that morning. And while the crowds of spectators and technicians paced the ground and chewed their fingernails, they all asked one question of each other: "How do you suppose he feels?" Some of the observers decided that Al's situation was perhaps a little easier than anyone else's. He had made his decision and he was committed. Now he had only to wait while other people made their decisions, and these were not so simple. A man's life was in their hands, and it was not exactly an enviable position that morning to be a test director or a flight director— or even, for that matter, a spectator.

This editor was on hand at the press site that morning, a mile or so from Pad 5. Reporters and TV cameramen had come out long before sunrise in Air Force buses and had been trying to steel their own nerves for a number of hours with paper cups full of coffee doled out from a mobile kitchen. At 6:27 they could all hear four mournful blasts from what sounded like a foghorn echoing across the palmetto flats which separated them from the pad. This was the standard warning signal that a missile was about to take off and that safety precautions should be observed. Then, seven minutes later, as the gantry moved away, the journalists could see the Redstone itself, standing alone and ready against the sky. On top of it was the tiny capsule, and on top of that was the orange-painted escape tower which would pull the capsule away from the booster in case of trouble. The reporters watched as a gangling cherry-picker crane, also painted a bright orange, edged up to the Redstone and nuzzled a boxlike cab just outside Shepard's hatch. It provided another way to carry Shepard to safety if the booster beneath him should threaten to blow up. Two journalists stood watching all of this activity together. They both knew Shepard well, and as the point finally got across to them that the moment of truth had arrived and that a man with twinkling eyes, a lovely wife and a white sports car was about to blast away at a velocity of 7,400 feet per second, one of the reporters decided that he could not bear to move his eyes from the capsule for a single second. His colleague, on the other hand, could not bear to watch it.

The long holds were excruciating—again, perhaps, for nearly everyone on the Cape but Alan Shepard. There were four of them in all and the first one began at 7:14 when the count had progressed with apparent ease but with increasing suspense until finally it stood at the fairly climactic reading of T minus 15 minutes. By this time, however, a thick muggy layer of clouds had begun to drift in over the launch site, and a hold was called to give the control center an opportunity to check on the weather. Cape Canaveral sits on a narrow spit of land with the Gulf Stream close by to the east and the Gulf of Mexico only 130 miles to the west. The weather is likely to shift rapidly between these two bodies of water. The day can be bright and sunny one minute, cloudy and breezy the next. It is fickle and difficult to keep track of, but the Control Center needed a firm prediction. The technicians knew that the weather downrange, where the recovery ships required perfect visibility to find the capsule and pick up Shepard, was excellent. But in order to follow the capsule and the booster closely, photograph their performance and watch for possible emergencies, the men in the Control Center required a clear view of the first part of the flight. The sudden appearance of small clouds blowing over the Cape could delay the entire mission. The appearance of a small hole in the clouds, however, gave Walt Williams, the operations director, enough confidence to carry on. The meteorologists reported that the clouds would soon blow away and that the sky would be clear again within about 30 minutes. Walt decided to recycle the count—or set it back—to allow for this delay, and he let the mission proceed.

Then another problem cropped up. During the delay for weather, a

small inverter which was located near the top of the Redstone began to
overheat. Inverters are used to convert the direct current which comes
from the batteries into the alternating current which is required to power
some of the systems in the booster. This particular inverter provided
400-cycle power which was needed to get the missile into operation and
off the pad. It had to be replaced before the mission could continue. This
involved bringing the gantry back into position around the Redstone
so that the technicians could get at it. The work was so time-consuming
that a total of 86 minutes went by before the Control Center could re-
sume the count.

I continued to feel fine. The doctors could tell how I was doing by
looking at their instrument panels and talking it over with me. I chatted
with Bill Douglas and Gordon Cooper and Paul Donnelly, the NASA
test conductor, during this period. I asked Deke Slayton to ask Shorty
Powers to call Louise and explain the reason for the holds and tell her
that I was fine. I also talked for a minute with Dr. Wernher von Braun,
whose people out at Huntsville had built the Redstone I was sitting on.

When the inverter was fixed, the gantry moved away, the cherrypicker
maneuvered its cab back outside the capsule, and the count went along
smoothly for 21 minutes before it suddenly stopped again. This time the
technicians wanted to double-check a computer which would help predict the trajectory of the capsule and its impact point in the recovery area.

I think that my basic metabolism started to speed up along about this
point. Everything—pulse rate, CO_2 production, blood pressure—began
to climb. I suppose that my adrenalin was flowing pretty fast, too. I was
not really aware of it at the time. But once or twice I had to warn myself, "You're building up too fast. Slow down. Relax." Whenever I
thought that my heart was palpitating a little, I would try to stop whatever I was doing and look out through the periscope at the people or at
the waves along the beach before I went back to work.

The last hold came at T minus 2 minutes and 40 seconds. This time
the people in the blockhouse were worried about the pressure on the
supply of lox in the Redstone which would be needed to feed the
rocket engines. The pressure on the fuel read 100 pounds per square inch
too high on the blockhouse gauge, and if this meant resetting the pressure
valve inside the booster manually, the mission would have to be scrubbed
for at least another 48 hours. Fortunately, the technicians found they

were able to bleed off the excess pressure by turning some of the valves by remote control, and the final count resumed at 9:25. Al had become slightly and understandably impatient at this point. He had been locked inside the capsule for nearly four hours, and as he listened to the worried engineers chattering to one another over the radio and debating about whether or not to repair the trouble, he got the impression that they were being too cautious just for his sake and might wind up taking too long. So he went on the intercom system with his only terse remark of the day.

"I'm cooler than you are," he said into the mike. "Why don't you fix your little problem and light this candle?"

The engineers made their fix and the orange cherrypicker moved away from the capsule for the last time. By now the television networks had been allowed to go on the air with a live broadcast of the proceedings. And across the nation millions of Americans became completely and emotionally involved with a man whom they did not know but whose feelings they tried to share in what was undoubtedly the greatest mass demonstration of empathy and the first coast-to-coast case of butterflies the nation had ever experienced.

On the beaches nearby the crowd could hear the countdown now on their transistor radios as they continued to stare at the booster. From T minus 10 minutes, when the broadcasts began, no one moved and no one spoke. The single exception, perhaps, was a lady who was quietly describing the lift-off to a friend of hers, who was blind.

"I don't see too well," Mrs. Charlotte Longo had told the people around her as she arrived. "But I wouldn't miss this shot."

At 9:34 EST on May 5, 1961, the technicians got to the bottom of their long countdown list and there was nothing left to do but push the button which ignited the Redstone. The flames roared out of the engine and licked across the pad, searing the concrete. For a second, the Redstone seemed to hesitate. Then, slowly and agonizingly at first, it began to climb. In the Mercury Control Center an elapsed-time indicator mounted behind the big status board began to thump off the passing seconds. A "time-to-retro-fire" clock began to click off in reverse, showing the seconds still remaining before the rockets should fire. On a wire stretched across the big map, a small symbolic capsule which marked the flight path for the observers began to quiver and start its puppet-like run out over the Atlantic. These were all visual clues to help the Mercury staff follow Shepard's progress. They all indicated that the mission was

*going fine. But the best clue of all was the voice of Alan Shepard as it
came back from the capsule by radio and echoed through the headsets.*

"Roger, lift-off and the clock is started," he shouted.

"Reading you loud and clear."

*"This is Freedom Seven. The fuel is 'Go.' One-point-two G. Cabin
at fourteen psi. Oxygen is 'Go.'"*

The last few minutes went perfectly. Everyone was prompt in his
reports. I could feel that all of the training we had gone through with the
blockhouse crew and booster crew was really paying off down there.
I had no concern at all. I knew how things were supposed to go, and that
is how they went. About three minutes before lift-off, the blockhouse
turned off the outside flow of cooling freon gas—I knew it would shut
off anyway at T minus 35 seconds when the umbilical fell away. At two
minutes before launch, I set the control valves for the suit and cabin
temperature, shifted to the voice circuit and had a quick radio check with
Deke. I also contacted Chase One and Chase Two—Wally and Scott
in the chase planes—and heard them loud and clear. They were in the
air, ready to take a high-level look at me as I went past them after the
launch. Electronically speaking, my colleagues were all around me at
this moment.

Deke gave me the count at T minus 90 seconds and again at T minus
60. I had nothing to do just then but maintain my communications, so
I rogered for both messages. At 35 seconds I watched through the periscope as the umbilical which had fed freon and power into the capsule
snapped out and fell away. Then the periscope came in and the little
door which protected it in flight closed shut. The red light on my instrument panel went out to signal this event, which was the last critical
function the capsule had to perform automatically before we were ready
to go. I reported this to Deke, and then I reported the power readings.
Both were in a "Go" condition. I heard Deke roger for my message, and
then I listened as he read the final count. "Ten, nine, eight, seven . . ."
At the count of 5 I put my right hand on the stop-watch button, which
I had to push at lift-off to time the flight. I put my left hand on the abort
handle, which I would move in a hurry only if something went seriously
wrong and I had to activate the escape tower.

Just after the count of zero, Deke said, "Lift-off." Then he added a
final tension-breaker to make me relax.

"You're on your way, José," he said.

I think I braced myself a bit too much while Deke was giving me the final count. Nobody knew, of course, how much chock and vibration I would really feel when I took off. There was no one around who had tried it and could tell me; and we had not heard from Moscow how it felt. I was probably a little too tense. But I was really exhilarated and pleasantly surprised when I answered, "Lift-off and the clock is started."

There was a lot less vibration and noise rumble than I had expected. It was extremely smooth—a subtle, gentle, gradual rise off the ground. There was nothing rough or abrupt about it. But there was no question that I was going, either. I could see it on the instruments, hear it on the headphones, feel it all around me.

It was a strange and exciting sensation. And yet it was so mild and easy —much like the rides we had experienced in our trainers—that it somehow seemed very familiar. I felt as if I had experienced the whole thing before. I knew, of course, that I had not. Nothing could possibly simulate in every detail the real thing that I was going through at that moment, and I tried very hard to figure out all of the sensations and to pin them down in my mind in words which I could use later. I knew that the people back on the ground—the engineers, doctors and psychiatrists—would be very curious about how I was affected by each sensation and that they would ask me quite a lot of questions when I got back. I tried to anticipate these questions and have some answers ready.

For the first minute, the ride continued to be very smooth. My main job just then was to keep the people on the ground as relaxed and informed as possible. It was no good for them to have a test pilot up there unless they knew fairly precisely what he was doing, what he saw and how he felt every 30 seconds or so along the way. So I did quite a bit of reporting over the radio about oxygen pressure and fuel consumption and cabin temperature and how the Gs were mounting slowly, just as we had predicted they would. I do not imagine that future spacemen will have to bother quite so much about some of these items. This was the first time, so we were being cautious.

I was scheduled to communicate about something or other for a total of 78 times during the 15 minutes that I was up. And I had to manage or at least monitor a total of 27 major events in the capsule. This kept me rather busy. But we wanted to get our money's worth when we planned this flight, and we filled the flight plan and the schedule with all the things we wanted to do and learn. We rigged two movie cameras inside the capsule, for example, one of which was focused on the instrument panel to keep a running record of how the system behaved. The other one was aimed at me to see how I reacted. We wanted a visual record of how my head and eyes moved as each event rolled by. The camera worked fine, and when we ran the film off the scientists were able to compile a chart of all my eye movements, which they related to the position of the instruments I had to watch as each moment and event transpired. On the basis of this data they later moved a couple of the instruments closer together on the panel so that future passengers would not have to move their eyes so often to keep up with things.

One minute after lift-off the ride did get a little rough. This was where the booster and the capsule passed from sonic to supersonic speed and then immediately went slicing through a zone of maximum dynamic pressure as the forces of speed and air density combined at their peak. The spacecraft started vibrating here. Although my vision was blurred for a few seconds, I had no trouble seeing the instrument panel. I decided not to report this sensation just then. We had known that something like this was going to happen, and if I had sent down a garbled message that it was worse than we had expected and that I was really getting buffeted, I think I might have put everybody on the ground into a state of shock. I did not want to panic anyone into ordering me to leave. And I did not *want* to leave. So I waited until the vibration stopped and let the Control Center know indirectly by reporting to Deke that it was "a lot smoother now, a lot smoother."

The pressure in the cabin held at 5.5 pounds per square inch, just as it was designed to do. And at two minutes after launch, at an altitude of about 22 miles, the Gs were building up and I was climbing at a speed of 3,200 miles per hour. The ride was fine now, and I made my last transmission before the booster engine cut off: ". . . All systems are 'Go.' "

The engine cutoff occurred right on schedule, at two minutes and 22 seconds after lift-off. Nothing abrupt happened, just a delicate and gradual dropping off of the thrust as the fuel flow decreased. I heard a roaring noise as the escape tower blew off. I was glad I would not be needing it any longer. I had hoped I could see smoke from the explosions blow past the portholes when this happened, but I was too busy keeping track of various events on the instrument panel to take a look. I reported all of these events to Deke, and then I heard a noise as the little rockets fired to separate the capsule from the booster. This was a critical point of the flight, both technically and psychologically. I knew that if the capsule got hung up on the booster, I would have quite a different flight, and I had thought about this possibility quite a lot before lift-off. There is good medical evidence to the effect that I was worried about it again when it was time for the event to take place, for my pulse rate reached its peak here—138. It started down again right away, however. (About one minute before lift-off my pulse was 90, and Gus told me later that when he and John Glenn saw this on the medical panel in the Control Center, they figured that my pulse was a good six points lower than Gus thought his was and eight points lower than John's.)

Right after leaving the booster, the capsule and I went weightless together and I could feel the capsule begin its slow, lazy turnaround to get into position for the rest of the flight. It turned 180 degrees, with the blunt or bottom end swinging forward now to take up the heat. It had been facing down and backwards. The periscope went back out again at this point, and I was supposed to do three things in order: (1) take over manual control of the capsule; (2) tell the people downstairs how the controls were working; and (3) take a look outside to see what the view was like.

The capsule was traveling at about 5,000 miles per hour now, and up to this point it had been on automatic pilot. I switched over to the manual control stick, and tried out the pitch, yaw and roll axes in that order. Each time I moved the stick, the little jets of hydrogen peroxide rushed through the nozzles on the outside of the capsule and pushed it or twisted it the way I wanted it to go. When the nozzles were on at full blast, I could hear them spurting away over the background noise in my headset. I found out that I could easily use the pitch axis to raise or lower the blunt end of the capsule. This movement was very smooth and precise, just as it had been on our ALFA trainer. I fed the yaw axis, and this maneuver worked, too. I could make the capsule twist slightly from left to right and back again, just as I wanted it to. Finally I took over control of the roll motion and I was flying Freedom 7 on my own. This was a big moment for me, for it proved that our control system was sound and that it worked under real space-flight conditions.

It was now time to go to the periscope. I had been well briefed on what to expect, and one of the last things I had done at Hangar S before suiting up was to study, with Bill Douglas and John Glenn, some special maps which showed me the view I would get. I had some idea of the huge variety of color and land masses and cloud cover which I would see from 100 miles up. But no one could be briefed well enough to be completely prepared for the astonishing view that I got. My exclamation back to

Deke about the "beautiful sight" was completely spontaneous. It was breath-taking. To the south I could see where the cloud cover stopped at about Fort Lauderdale, and that the weather was clear all the way down past the Florida Keys. To the north I could see up the coast of the Carolinas to where the clouds just obscured Cape Hatteras. Across Florida to the west I could spot Lake Okeechobee, Tampa Bay and even Pensacola. Because there were some scattered clouds far beneath me I was not able to see some of the Bahama Islands that I had been briefed to look for. So I shifted to an open area and identified Andros Island and Bimini. The colors around these ocean islands were brilliantly clear, and I could see sharp variations between the blue of deep water and the light green of the shoal areas near the reefs. It was really stunning.

But I did not just admire the view. I found that I could actually use it to help keep the capsule in the proper attitude. By looking through the periscope and focusing down on Cape Canaveral as the zero reference point for the yaw control axis, I discovered that this system would provide a fine backup in case the instruments and the autopilot happened to go out together on some future flight. It was good to know that we could count on handling the capsule this extra way—provided, of course, that we had a clear view and knew exactly what we were looking at. Fortunately, I could look back and see the Cape very clearly. It was a fine reference.

All through this period, the capsule and I remained weightless. And though we had had a lot of free advice on how this would feel—some of it rather dire—the sensation was just what I expected it would be: pleasant and relaxing. It had absolutely no effect on my movements or my efficiency. I was completely comfortable, and it was something of a relief not to feel the pressure and weight of my body against the couch. The ends of my straps floated around a little, and there was some dust drifting around in the cockpit with me. But these were unimportant and peripheral indications that I was at zero G. As for the dust, we always tried to keep the capsule sterile, of course. But the only way to do that completely is just not to let anybody get into it—not even a pilot. On future flights, when we will want to open up our faceplates for quite a while during a long ride, the business of having dust float around in front of us may get serious. But we will face that when we get to it and find a solution.

At about 115 miles up—very near the apogee of my flight—Deke Slayton started to give me the countdown for the retro-firing maneuver. This had nothing directly to do with my flight from a technical standpoint.

I was established on a ballistic path and there was nothing the retro-rockets could do to sway me from it. But we would be using these rockets as brakes on the big orbital flights to start the capsule back towards earth. We wanted to try them on my trip just to see how well they worked. We also wanted to test *my* reactions to them and check on the pilot's ability to keep the capsule under control as they went off. I used the manual control stick to tilt the blunt end of the capsule up to an angle of 34 degrees above the horizontal—the correct attitude for getting the most out of the retros on an orbital re-entry. At five minutes and 14 seconds after launch, the first of the three rockets went off, right on schedule. The other two went off at the prescribed five-second intervals. There was a small upsetting motion as our speed was reduced, and I was pushed back into the couch a bit by the sudden change in Gs. But each time the capsule started to get pushed out of its proper angle by one of the retros going off I found that I could bring it back again with no trouble at all. I was able to stay on top of the flight by using the manual controls, and this was perhaps the most encouraging product of the entire mission.

Another item on my schedule was to throw a switch to try out an ingenious system the engineers had worked out for controlling the attitude of the capsule in case the automatic pilot went out of action or we were running low on fuel in the manual control system. We have two different ways of controlling the attitude of the capsule—manually with the control stick, or electrically with the autopilot. In the manual system the movement of the stick activates valves which squirt the hydrogen peroxide fuel out to move the capsule around and correct its attitude. We can control the magnitude of this correction by the amount of pressure we put on the stick. If we give it just a little flick, we move the capsule just a hair. If we give it a real twist with the wrist, we can move the capsule quite a bit. The choice depends on how far off we think we are. The autopilot works differently. It uses an entirely different set of jets —to give us a backup capability in case one set goes out—and a separate source of fuel. There is no variation of magnitude in the automatic jets. The autopilot activates electric solenoids which open and close the valves. They either squirt or they don't. There are two sizes of jets in both systems, for small thrust and large thrust. But the automatic jets are not proportional in the force that they exert. This gave the engineers an idea: they created a third possibility, which they call "fly-by-wire," in which the pilot switches off the automatic pilot, then links up his manual stick with the valves that are normally attached to the automatic system. This gives him a new source of fuel to tap if he is running low, and a little more flexibility in managing the controls. The fly-by-wire mode seemed fine as far as I was concerned, and another test was checked off the list of things we were out to prove.

We were on our way down now and I waited for the package which holds the retro-rockets on the bottom of the capsule to jettison and get out of the way before we began our re-entry. It blew off on schedule and I could feel it go. I watched through the periscope as some of the straps which had held it in place fell away. The retro-package passed me again as it fell back towards the ocean, but the green light which was supposed to report this event failed to light up on the instrument panel. This was our only signal failure of the mission. I pushed an over-ride button, however—which bypasses the automatic signaling system and straightens things out manually—and the light turned green as it was supposed to do. This meant that everything was all right.

Now I began to get the capsule ready for re-entry. Using the control stick, I pointed the blunt end downward at about a 40-degree angle, and switched the controls back to the autopilot so I could be free to take another look through the periscope. The view was still spectacular. The sky was very dark blue; the clouds were a brilliant white. Between me and the clouds was something murky and hazy which I knew to be the refraction of the various layers of the atmosphere through which I would soon be passing.

I fell slightly behind in my schedule at this point. I was at about 230,000 feet when I suddenly noticed a relay come on which had been activated by a device that measures a change in gravity of .05 G. This was the signal that the re-entry phase had begun. I had planned to be on manual control when this happened and run off a few more tests with my hand controls before we penetrated too deeply into the atmosphere. But the G forces had built up before I was ready for them, and I was a few seconds behind. I was fairly busy for a moment running around the cockpit with my hands, changing from the autopilot to manual controls, and I managed to get in only a few more corrections in attitude. Then the pressure of the air we were coming into began to overcome the force of the control jets, and it was no longer possible to make the capsule respond. Fortunately, we were in good shape, and I had nothing to worry about so far as the capsule's attitude was concerned. I knew, however, that the ride down was not one most people would want to try in an amusement park.

In that long plunge back to earth, I was pushed back into the couch with a force of about 11 Gs. This was not as high as the Gs we had all taken on the centrifuge at Johnsville during the training program, and I remember being clear all the way through the re-entry phase. I was able to report the G level with normal voice procedure, and I never reached the point—as I often had on the centrifuge—where I had to exert the maximum amount of effort to speak or even breathe. All the way down, as the altimeter spun through mile after mile of descent, I kept grunting out "O.K., O.K., O.K.," just to show them back in the Control Center how I was doing. The periscope had come back in automatically before the re-entry started. And there was nothing for me to do now but just sit there, watching the gauges and waiting for the final act to begin.

All through this period of falling the capsule rolled around very slowly in a counterclockwise direction, spinning at a rate of about 10 degrees per second around its long axis. This was programed to even out the heat and it did not bother me. Neither did the sudden rise in temperature as the friction of the air began to build up outside the capsule. The temperature climbed to 1230 degrees Fahrenheit on the outer walls. But it never went above 100 degrees in the cabin or above 82 degrees in my suit. The life-support system which Wally had worked on—oxygen, water coolers, ventilators and suit—were all working without a hitch. As the G forces began to drop off at about 80,000 feet, I switched back to the autopilot again. By the time I had fallen to 30,000 feet the capsule had slowed down to about 300 miles per hour. I knew from talking to Deke that my trajectory looked good and that Freedom 7 was going to land right in the center of the recovery area. But there were still several things that had to happen before I could stretch out and take it easy. I began to concentrate now on the parachutes. The periscope jutted out again at about 21,000 feet, and the first thing I saw against the sky as I looked through it was the little drogue chute which had popped out to stabilize my fall. So far, so good. Then, at 15,000 feet, a ventilation valve opened up on schedule to let cool fresh air come into the capsule. The next thing I had to sweat out was the big 63-foot main chute, which was due to break out at 10,000 feet. If it failed to show up on schedule I could switch to a reserve chute of the same size by pulling a ring near the instrument panel. I must admit that my finger was poised right on that ring as we passed through the 10,000-foot mark. But I did not have to pull it. Looking through the periscope, I could see the antenna canister blow free on top of the capsule. Then the drogue chute went floating away, pulling the canister behind it. The canister, in turn, pulled out the bag which held the main chute and pulled *it* free. And then, all of a sudden, after this beautiful sequence, there it was—the main chute stretching out long and thin. It had not opened up yet against the sky. But four seconds later the reefing broke free and the huge orange and white canopy blossomed out above me. It looked wonderful right from the beginning. I stared at it hard through the periscope for any signs of trouble. But it was drawing perfectly, and a glance at my rate-of-descent indicator on the panel showed that I had a good chute. It was letting me down at just the right speed and I felt very much relieved. I would have a nice, easy landing.

The water landing was all that remained now, and I started getting set for it. I opened the visor in the helmet and disconnected the hose that keeps the visor sealed when the suit is pressurized. I took off my knee straps and released the strap that went across my chest. The capsule was swaying gently now, back and forth under the chute. I knew that the people back in the Control Center were anxious about all this, so I sent two messages—one through a voice-relay airplane which was hovering around nearby, and the other through a telemetry ship which was parked in the recovery area down below. Both messages read the same: "All O.K."

At about 1,000 feet I looked out through the porthole and saw the water coming up towards me. I braced myself in the couch for the impact, but it was not at all bad. It was a little abrupt, but no more severe than the jolt a Navy pilot gets when he is launched off the catapult of a carrier. The spacecraft hit and then it flopped over on its side so that I was leaning over on my right side in the couch. One porthole was completely under water. I hit the switch to kick the reserve parachute loose. This would take some of the weight off the top of the capsule and help it right itself. The same switch started a sequence which deployed a radio antenna to help me signal my position. I could see the yellow dye marker coloring the water through the other porthole. This meant that the other recovery aids were working. I could not see any water seeping into the capsule, but I could hear all kinds of gurgling sounds around me, so I was not sure whether we were leaking or not. I remember reassuring myself that I had practiced getting out of the capsule under water and that I could do it now if I had to. But I did not have to try. Slowly but steadily the capsule began to right itself. As soon as I knew the radio antenna was out of the water I sent off a message saying that I was fine.

I took off my lap belt and loosened my helmet so I could take it off quickly when I went out the door. And I had just started to make a final reading on all of the instruments when the helicopter pilot called me. I had already told him that I was in good shape, but he seemed in a hurry to get me out. I heard the shepherd's hook catch hold of the top of the capsule, and then the pilot called again.

"O.K.," he said, "you've got two minutes to come out."

I decided he knew what he was doing and that following his instructions was perhaps more important than taking those extra readings. I could still see water out of the window, and I wanted to avoid getting any of it in the capsule, so I called the pilot back and asked him if he would please lift the capsule a little higher. He obligingly hoisted it up a foot or two. I told him then that I would be out in 30 seconds.

I took off my helmet, disconnected the communications wiring which linked me to the radio set and took a last look around the capsule. Then I opened the door and crawled to a sitting position on the sill. The pilot lowered the horse-collar sling; I grabbed it, slipped it on and then began the slow ride up into the helicopter. I sank into a bucket seat as soon as I reached the top, and on the way to the carrier I felt relieved and happy. I knew I had done a pretty good job. The Mercury flight systems had worked out even better than we had thought they would. And we had put on a good demonstration of our capability right out in the open where the whole world could watch us taking our chances.

It took the helicopter seven minutes to get me to the carrier. When we approached the ship, I could see sailors crowding the deck, applauding and cheering and waving their caps. I felt a real lump in my throat. I started for the quarters where the doctors would give me a quick once-over before I flew on to Grand Bahama Island for a full debriefing. First, however, I went back to the capsule, which had been lowered gently onto a pile of mattresses on the carrier deck. I wanted to retrieve the helmet I had left behind in the cockpit. And I wanted to take one more look at Freedom 7. I was pretty proud of the job that *it* had done, too.

When all of the returns of Shepard's flight were in, when all of the debriefings were over and the volumes of data had been analyzed, it was clear that Al Shepard had completed the most grueling ride any American had ever taken. The capsule carried him 302 miles downrange in 15 minutes and took him to a maximum altitude of 116.5 statute miles—an apogee which he reached at T plus 5 minutes and 11 seconds. His maximum velocity—which came at T plus 2 minutes and 22 seconds—was 7,388 feet per second, or 5,036 miles an hour. At T plus 8 minutes and 20 seconds he was pulling 11 Gs. Twenty seconds later, before he started his parachute drop of 30 feet per second, he was living once more in man's normal environment of 1 G. He bounced up briefly to 3 or 4 Gs when the main chute opened and decelerated his fall. Then, 5 minutes and 7 seconds later, he was back to 1 G again and resting in the water.

The capsule came through the entire flight in such excellent shape that the engineers who went over it with a fine-tooth comb decided that it could easily be used again.

The doctors, who went over Alan Shepard with everything but a comb, decided that the commander could be used again, too. The psychiatrists, who had had their doubts about what the effects of such physical and mental stress and tension might be on a man, sat Al down for some long talks when he came back. They then made their report.

"Subject felt calm and self-possessed," they wrote with unconcealed admiration. "Some degree of excitement and exhilaration was noted. He was unusually cheerful and expressed delight that his performance during the flight had actually been better than he had expected. . . . He was more concerned about performing efficiently than about external dangers. He reported moderate apprehension during the preflight period, which was consciously controlled by focusing his thoughts on technical details of his job. As a result, he felt very little anxiety during the immediate prelaunch period. After launch he was preoccupied with his duties and felt concern only when he fell behind on one of his tasks. (This was the manual control at re-entry.) There were no unusual sensations regarding weightlessness, isolation or separation from the earth."

The physicians had their turn with Shepard, too, and made a similarly positive report. They found, among other things, that the flow of norephinephrine from Shepard's adrenal glands during the flight had been more than two and a half times what it had normally been. He had maintained a pulse rate of about 80 beats per minute during the countdown. His normal pulse was 65. The pulse rate rose briefly to 90 and 95 during various significant checkout events of the countdown and immediately after the hatch was locked on. It rose to 108 beats per minute just 30 seconds prior to lift-off, and stood at 126 at the lift-off signal when the count had reached zero. His pulse reached its peak at 138 when the Redstone engine cut off, and it remained there for approximately 45 seconds. It fluctuated between 108 and 130 beats per minutes during his parachute descent.

Shepard's respiration rate behaved in similar fashion. He was taking approximately 15 to 20 breaths per minute during the countdown—his normal rate of breath is about 10. His breathing speeded up to 30-35 breaths per minute during launch, stood at 30 during re-entry, and fluctuated between 20 and 25 breaths per minute during the descent.

Shepard's deep-body temperature was 98.4 degrees Fahrenheit when he climbed into the capsule and it rose to a peak of 99.2 at a point near the end of the flight when his entire body was being warmed by the heat of re-entry. He lost three pounds during the mission. He weighed 169 pounds and 4 ounces before the final countdown began and was down to 166 pounds and 4 ounces when the Navy doctors weighed him aboard the carrier. This loss was not considered unusual and was ascribed to the fact that he perspired a good deal in his hot rubberized spacesuit as he was being carried back to the ship. He had simply sweated it off.

Through all of the probings and measurements to which Shepard had to submit in order to gather such facts as these for the medical books, he was a patient and understanding guinea pig. He had only one suggestion to make when it was all over. He wondered, he said, if the doctors really needed to drain from him all of the body-fluid samples which they took at various times both before and after the flight.

"I felt," he said, looking around the room with a wry smile, "as though an unusual number of needles were used."

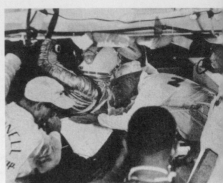

was shortly after 5 on the morning of May
1961, when Commander Alan B. Shepard,
., walked down the steps of the van that
d taken him to the launching pad and got
s first look at the steaming, white Redstone
hich would launch him into space. He rode
e elevator to the top of the gantry, and a
w minutes later, with John Glenn carefully
uiding him over the hatch so he would not
ar his space suit, Shepard squeezed into the
apsule [TOP, LEFT].

Back at Mercury Control, a few miles
way, Flight Director Chris Kraft stood at
s console with microphone and earphones
place and went through the final check-
ut procedures as Operations Director
Walter Williams looked on from his console
the rear of the big, glassed-in room [CEN-
ER, LEFT].

And then, at 9:34 EST, the countdown
ached 0. The slender Redstone blasted
raight up from its pad, and a crowd of tense
d prayerful people, some of whom had
eped out to a remote vantage point,
atched the Redstone soar and disappear
to the sun, leaving behind a squiggly white
ontrail to show where it had passed.

The flight, which lasted 15 minutes, was smooth and went according to plan. Astronaut Wally Schirra monitored it from his post at a tracking station on a rocky beach in Bermuda.

Shepard's capsule had no sooner hit the water, 302 miles from Canaveral, than a helicopter flew in overhead, hooked onto the capsule to lift it out of the water,

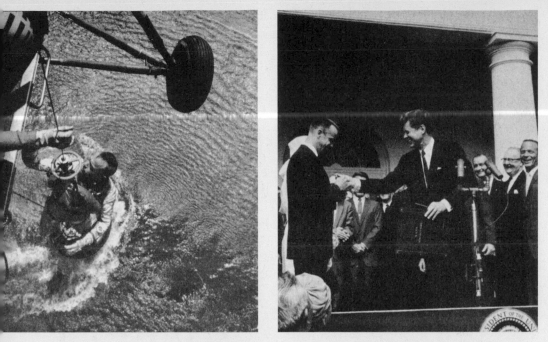

lowered a sling for Shepard himself, and then, when Shepard had been hoisted inside, flew both to a waiting aircraft carrier. A few days later Shepard accepted a medal from President Kennedy at the White House, and then they went inside where congressmen, NASA officials and other Astronauts listened as Shepard told the President how it had gone.

THE
WITH

Virgil I. Grissom

TROUBLE

LIBERTY BELL

Actually, during the final weeks and days before the launching of MR-4—as my flight was labeled—I really felt good. We kept spotting problems, as we knew we would. But there were very few of them, considering the state of the art, and the simulations we went through for practice went very well. If anything was building up inside me, it was that I was anxious. I kept wanting to go tomorrow, and I guess I got slightly impatient whenever some technician came up with a new modification in the system that might have caused a long delay if we had accepted it. The only thing I was afraid of was that something might happen to prevent *me* from making the flight. So I gave up water skiing and I was more careful than usual to observe the speed limits around the Cape.

I had my first look at Capsule 11, which was assigned to me, at the McDonnell plant in St. Louis in January, about six months before my flight was scheduled. It arrived at Hangar S at the Cape on March 7, and we had it ready for launching 136 days later. There was a lot of work to be done right there at the Cape, but I did not get to spend much time with my spacecraft until after Al Shepard's flight. We all pulled together on that one, just as we do on all our flights, and I stuck pretty close to Al until he went up. I was with him when he dressed, and I rode out in the van with him to the pad. After that flight I buckled down to my own problems and stayed near my own capsule as much as I could. I had the first picture window to look through, for example, which meant that I needed some extra training in visual navigation. I also had a new gimmick in the manual control system. Mine was rigged so that I could control not only the attitude of the capsule but also the rate at which the corrections were made by the nozzles. This required some extra practice. And I had the new hatch which Deke Slayton has described, the lighter model with seventy explosive bolts to hold it in place.

Things moved so fast, however, that I did not get to do quite all of the things I wanted to do before the flight. I would have liked to practice more often on the ALFA trainer I have described earlier, which allows you to lie back on the couch and practice maneuvering yourself with respect to the earth and the horizon outside. I had a total of thirty-six missions on ALFA, but I felt that I could use more—especially with the new window in place so I could become more familiar with the technique. But the only ALFA trainer we had was installed back at our headquarters at Langley. And once I got immersed in all of the work at the Cape, I just did not have the chance to get back to Langley for all of the additional workouts that I wanted. I also feel—now that it is all over—that I could have used more training in recovery procedures at sea. But this is completely a matter of hindsight. At the time, I felt that I was well prepared. I do value the time that I spent on the centrifuge. I went through fifteen different missions on the "Wheel," simulating the same flight profile that I would on the Redstone. These sessions prepared me very well for the sensations I ran into later on the actual flight. And I spent one hundred very productive hours in the procedures trainer, practicing over and over again every conceivable combination of button pushing, lever pulling, toggle switching and emergency procedures that I might have to resort to in a hurry on the flight itself.

Before I got immersed in this final phase of training, however, I spent most of my time worrying about my capsule. I made a point of attending production meetings at the McDonnell plant in St. Louis, supervising some of the engineering work periods and fretting a little over whether all of the critical parts would arrive from the subcontractors on time and get put together. I thought it would be good for the engineers and workmen who were building my spacecraft to see the pilot who would have to fly it hanging around. It might make them just a little more careful than they already were and a little more eager to get the work done on time if they saw how much I cared.

We had to make one really important fix in the capsule early in the game. The controls got snarled up somehow in the assembly at the plant, and the first time I tested them out the system yawed to the left when it should have gone to the right. This sort of thing should not happen even on automobiles. But the mistake was soon corrected. We had another problem that we never did get fixed. The attitude controls failed to center themselves again after a yawing maneuver, and I had to compensate for this myself. So long as I knew about this idiosyncrasy ahead of time and could make allowances for it, there was no trouble. I could live with it.

It took a total of 33 days of work, after the spacecraft arrived at the Cape, to take some of it apart again, test each component thoroughly, put it back together again, and then test the entire system again after everything had been reinstalled. I lay in my couch for hours at a time, running off various tests. One of these, called SEDR 83 (SEDR stands for "Systems Engineering Department Report"), was concerned solely with checking out the environmental control system. Another was SEDR 61, which was a test of all the communications equipment in the spacecraft. SEDR 73 was a run-down of the reaction control system.

While all this was going on at Cape Canaveral, we also had daily scheduling meetings to keep everyone informed of our progress and up to date on any problems which cropped up. Here is where we reviewed the work being done on the various systems. It was also here that the perfectionists in the crowd would sometimes try to stop the show and redesign the whole system again from scratch. I wanted a safe and efficient capsule as much as anyone, and I did not blame the engineers, who were proud of their work, for trying to make each part they were working on absolutely perfect. I knew that if something happened to me which could be traced to one of their decisions, it would hang heavy on them. But I had also noticed, during my days as a test pilot, that engineers are seldom satisfied to have something work *well*. They often want to go on testing and testing a system until it is almost worn out. I felt, therefore, that it was up to me to stay on top of the situation and make sure that we got a spacecraft at all, and then try to reassure the engineers that if it satisfied the pilot who had to fly it, it ought to satisfy them. In order to convince them of my case, of course, I had to know the spacecraft myself, inside out.

My flight was scheduled for sometime in July. During the latter part of May, right after Al Shepard's mission, I had to start worrying about heading off some of the difficulties that he had run into. On Freedom 7, for example, the package holding the retro-rockets was jettisoned at the prescribed time, but the green light on Al's instrument panel which was supposed to tell him this failed to light up. Bob Foster, one of the top engineers with McDonnell, thought it might have happened because of an overload on the electrical circuit. So many things were going on at once that the batteries were not strong enough to handle them all. Bob had a couple of fixes in mind to repair this fault, but after a lot of discussion we decided to leave it alone and see what happened on my flight. The problem was not dangerous. The pilot could override it just as Al had. We also decided that if we tried to work out some short-cut solution we might never uncover the exact cause and could wind up with even more trouble. The problem of just when to accept this kind of calculated risk had been a matter of debate all through the program. Personally, I was usually in favor of taking the risk—provided it was not too serious. Foster also suggested a way to keep one of the inverters from overheating. This was the problem that caused so much trouble during the long hold that preceded Al Shepard's flight. It was more important than the failing green light, and we decided to buy the fix. It consisted of installing an extra switch on the fuse panel inside the capsule, which we would use just prior to launch.

During the first three weeks in June, while the capsule was still in the white room at Cape Canaveral waiting to be moved out to the launching pad, I kept an eye on it almost constantly except for a few brief quick trips back to our headquarters at Langley. We were working on a number of other problems there, and I wanted to be in on them. One was a new safety device to protect the recovery or rescue crews if they had to open up the capsule from the outside to get me out. The lanyard they would use to detonate the explosive bolts to blow the hatch off was only 42 inches long. But the explosive charges were powerful enough to throw the hatch about 20 feet. This meant that the rescue crews would be close enough to the capsule to get hurt trying to get me out. Dick Smith, one of our troops at Langley, came up with a solution. He fastened two soft iron coils onto the hatch which linked it to the capsule sill. When these were in place, the hatch could fly only so far. Then it would fall harmlessly down the side without knocking over any of the crew. Needless to say, I was all for this change.

I also had a chance to talk with Glenn Shewmake, our specialist in charge of personal equipment, about a new life raft which he had designed on his own and had built practically with his own hands. We liked the looks of it because it was a good deal lighter than the raft we had started out with. Wally Schirra and I tested it out in the bay at Langley, and it worked fine. I also made a run down to Pensacola, Florida, to have the raft evaluated by the Navy survival experts there. Unfortunately, the weather was bad when I arrived, and we had to scrub the test. We had our fishing poles along, but the fish weren't biting either. So I headed back to Canaveral.

In that same month we ran into a series of small but irritating difficulties which threatened to put us behind schedule. This possibility worried me, for though we had not fallen behind yet we had used up all of our extra contingency time, and any real bugs would set us back.

The first problem was a report from the Marshall Space Flight Center at Huntsville, Alabama, that the technicians were running into difficulty with our Redstone and might not be able to deliver it to the Cape by June 22, when we were supposed to get it. Then the technicians who were working on my pressure suit said they were not sure how to make a solid check of some of the minor items in the suit before they installed them. I suggested we put these items in unchecked, since I did not really need them anyway, and take a chance that they would work. It was my theory all along that we ought to have everything working perfectly that could possibly affect our safety or the efficiency of the capsule. But I was convinced that if we waited for all the peripheral equipment to work perfectly, too, we would never get off the ground. The engineers could not agree with my reasoning, and they went right on figuring.

Then I learned that the capsule clock, which keeps track of the elapsed time of the mission, had rusted out and we had no replacement on hand. Even though it was a standard eight-day aircraft clock, the supply people said that they could not get their hands on one before July 4. This was silly, and Sam Beddingfield, one of the NASA engineers, finally settled everything by promising he would produce a clock by the next morning —even if he had to steal one. I do not think Sam was a thief, but we got our clock. The other problems got solved too.

Things began to look up on June 22, when the Redstone was flown to the Cape in a big cargo plane, right on schedule. I went out to the airport to look it over, and when I got there the crew had already unloaded the bird and put it on a trailer for the slow ride to the launching pad about 15 miles away. I joined the caravan, and when we reached the pad I got out and walked along beside the Redstone as it pulled in. I guess I looked a little eager, for Paul Donnelly, the capsule test conductor, spoke up.

"Don't worry, Gus," he said, "they're not going to shoot it without you."

Everybody laughed, so I joined the fun and answered, "Hell, I've already ridden it bareback all the way from Huntsville."

On Sunday, June 25, we completed the final checkout of the capsule inside the hangar. This consisted of running off some simulated missions which lasted until early Monday morning. John Glenn, my backup pilot, was in the capsule for one of these final checkouts, and the two hours went smoothly except for two minor discrepancies. One was that the attitude for firing the retro-rockets was slightly out of limits. The other was a telemetry failure which prevented us from getting a reading on the temperature of the oxygen being fed into the pressure suit. We decided to repair these before going on. The second test went without a single hitch, and at midnight we began a test to see if the pilot could override and take over the controls and signals manually if for some reason the automatic system failed. This was a crucial test, but everything went fine. Then we simulated an abort situation to make sure that all of our emergency procedures would work in case of any trouble during launch. At 6:20 A.M., after a long, hard night, we wound up with a test of the switches. Then we had a meeting to review the results. We had to make a few minor fixes, but in general the capsule was now in good shape.

This was chiefly because we had some fine people slaving over it. We were all working around the clock during this period, running tests night and day in order to keep up with the schedule. Some of the men were on the edge of complete exhaustion. I remember John Shriefer, in particular. John was the McDonnell capsule chief from St. Louis, and one of the finest troops I ever met. I knew he had been knocking himself out getting the capsule ready and ironing out one problem after another. But I did not realize how worn out he was until one day we stopped for lunch and he was so exhausted he could not eat a bite. He just dipped his spoon into the soup. I took him right back to the Terrace Dune Apartments, where he was staying with his family, and asked his wife to keep him there and make him rest for twenty-four hours. We needed John, but if the only way to keep him on the job was to order him to bed, that was it. He came back to work the next day, rested and raring to go.

July 1 was a big day. This was when the capsule was taken from the hangar to the launch pad to be mated to the Redstone. It was the happiest moment to date for me. My machine was getting spliced together and we were right on schedule. I had only one problem that day: they almost wouldn't let me on the pad to keep an eye on things. I had locked my hard hat in the office and forgotten the key, and no one is allowed near an active gantry without a special hard hat to protect his head. Someone finally loaned me one, and I made it just in time.

From this day on, I was at the pad most of the time, either participating in or observing every test that was made on the capsule-booster combination. All told, it took us three weeks to test out the combination and make sure that the two big components were compatible. During this time I became familiar with the launch procedure and grew to know and respect the launch crews. I gained confidence in their professional approach and in the smooth way they executed the prelaunch tests.

But we still had some problems. Soon after we began the business of fitting the capsule to the Redstone and fastening them together with clamps and rings and connecting wires, I went to another scheduling meeting. Here I faced the threat of another delay. Someone wanted to change the position of a camera which was mounted in the cockpit. This would mean tearing up wires and changing relay panels and adding things, and I was against it. For one thing, I was afraid they might never get everything put back together in time. And for another, everything was in such good working shape that I did not want them tinkering with it. I clinched my argument by pointing out that the new position for the camera would make it more difficult for me to squeeze into the capsule—and, especially, to get out of it in a hurry. That did it. They dropped the idea.

On the fifth of July, a chartered plane brought Slayton, Schirra and Cooper to the Cape. This made me feel good. The scheduled launch was less than two weeks away and the other troops were moving into place. I went down to the launch pad the next morning and ran into Cooper, who showed me where the cherrypicker would be that would help get me out of the capsule in case of trouble during the countdown. Gordo had spent several weeks setting up emergency procedures—including fire-fighting crews and techniques for handling the volatile fuels and the various pyrotechnics like escape rockets and retro-rockets that had to be inserted into the system. He explained what the procedures would be in the case of each possible catastrophe. If the capsule and I toppled to the ground in a fire, for example, a special team of men including capsule technicians who knew where all of the seams and bolts were would rush into the area in fireproof suits to get me out. Dr. Dave Morris, the man who would be sweating out the launch in the cab of the cherrypicker, showed me his controls and explained how he would try to fish me out if I were still on top of the Redstone. I knew that an amphibious vehicle would be standing by with a skindiver, a medical man and more capsule technicians in case we had an abort on the pad and I got tossed into the ocean by the escape tower. I also knew that our team doctor, Bill Douglas, would be sitting in a helicopter during the countdown, along with another skindiver, and would be ready to fly to my rescue no matter where I wound up—in the water or on dry land. Bill would have a medical pack strapped to his back, complete with a supply of bandages, surgical tools and morphine. He and the helicopter would sweat out the launch next to a small, one-bed hospital near the launching pad where he had an operating room set up for emergencies. He also had a snake-bite kit on him. This was because the area around the pad where I might come plummeting down was crawling with rattlesnakes.

All of these facilities had been available for Al Shepard, too, but this was the first time that I had seen some of them. It was good to know that all of these fine people were ready, and that they had apparently thought of everything. They even had a backup helicopter ready, a backup amphibious launch, and a backup doctor in case Bill got held up.

I talked some more with Cooper and took a look at the safety device which would spring the hatch open in case of trouble. Then I rushed over to the hangar to call Betty. It was our wedding anniversary, and I was pretty sure it would be late that night before I would have another chance to call her. We talked for about fifteen minutes, and she seemed in good spirits.

On Friday, July 7, we ran a test to make sure that all of the radio frequencies we would be using during launch were compatible. If they were not—if one frequency was out of line and could interfere with another one—it was possible for a stray signal to activate the escape rocket or some of the other pyrotechnics and blow up the whole mission before I ever got my safety straps fastened. We went to a lot of trouble for this test. I suited up in the hangar, got into the van and rode out to the pad just as I would for the real thing. We picked up the countdown at T minus 170 minutes and turned on all the telemetry, radar and radio transmitters we would have running on launch day. I climbed into the capsule at T minus 123 minutes, right on schedule. Then they moved the gantry away, since the presence of all that steel next to the capsule would interfere with the test.

It was quite a sensation. There I was, perched up on top of that slender 70-foot booster with nothing to hold it upright and steady but a half-load of fuel in the tanks. I think a fairly strong wind could have pushed me over, otherwise. I looked out of the window as the gantry moved away, and for a moment I had quite a start. It looked as if the gantry were standing still and *I* were moving. I knew that I wasn't, but I got on the radio and said, "It looks like I'm falling." Bill Augerson, the NASA doctor who was on duty in the blockhouse, told me later that Al Shepard had experienced exactly the same sensation. Once I was up there all by myself and the gantry was gone, the movement of the capsule was considerably less than I had expected. There was a slight sway in the summer breeze, but the only thing I remember about the new view was some clouds floating peacefully by.

At the end of this check, we had another meeting and I passed out copies of my flight plan, which had changed very little since I first drew it up several months before. The plan differed slightly from Al's, for we wanted to find out some things that we had not tested on his flight. We weeded out a lot of the communications checks which Al had to worry about, for example, so I would have more time to make visual observations. I wanted to fly part of the mission visually rather than rely entirely on the instruments. I also wanted to give the manual control system a different workout. When Al switched from the automatic pilot to manual control he did it one axis at a time—pitch, yaw and roll. I planned to take over all three axes at once to give us another check on this system and see if I could handle it. I also planned to fire the retro-rockets manually instead of automatically as they had been fired on Freedom 7.

We had another hassle at this meeting. We had discovered a minor problem with the oxygen system before I left the capsule, and though I thought I had figured out a good way to solve it, some of the engineers seemed determined to start digging at it in more detail. I was against letting them tear the system apart and risk more delays trying to fix it. I guess I got rather impatient. It was a Friday evening, and this would be my last chance to fly home for a weekend with my family before the launch. So 45 minutes before my plane was due to leave, I got up and said, as firmly as I could, that I was satisfied the capsule was in good flying shape. "Please don't anybody fiddle with it over the weekend," I said. Then I went home.

I spent most of the weekend at home, although on Saturday morning I did drive over to Langley to get in some extra time on the ALFA trainer and practice manual control of the capsule. That afternoon, Betty and I went to pick up my oldest son, Scott, at Boy Scout camp. That night I took Betty out to dinner. Sunday morning, on my way back to the Cape, I drove the boys to Sunday School. It was then that I told them for the first time that if all went well I would be making a flight like Commander Shepard's.

Scott, who was eleven, said, "Gee, that's great, Dad."

But that was about it. I guess they have about as much emotional display as their father does. I did hear from Betty later that they were beaming from ear to ear when she picked them up after Sunday School. I had told them not to tell anyone about it yet, that it was a big secret. But Mark, who was seven, went over to Betty and whispered into her ear.

"Daddy's going," he said.

It made me feel good to know they were so happy about it.

I returned to the Cape that day and stayed there until launch. We ran off Tests Three, Four and Five the next day. All of them were simulated abort missions. We finished Three and Four with no trouble in the morning. We got the oxygen problem solved without tearing things apart. And though we discovered at the scheduling meeting that some of the fittings which clamped the capsule to the Redstone had become bent and would have to be replaced that night, we did not think this would hold us up. The technicians did have to break all of the electrical connections between the bird and the capsule to get this done, and then they had to mate the two all over again. Somehow, they accomplished the job in record time without leaving any loose ends.

One of the less vital problems I had was figuring out a name and an insignia for the capsule. As the pilot, I had the prerogative of thinking up a name. I decided on Liberty Bell, because the capsule does resemble a bell. John Glenn felt that the symbolic number "seven" should appear on all our capsules—in honor of the team—so this was added. Then one of the engineers got the bright idea that we ought to dress Liberty Bell up by painting a crack on it just like the crack on the real one. No one seemed quite sure what the crack looked like, so we copied it from the "tails" side of a fifty-cent piece. Ever since my flight—which ended up with the capsule sinking to the bottom of the Atlantic—there has been a joke around the Cape that that was the last capsule we would ever launch with a crack on it.

We kept running practice missions right up through the last week before the launch. On one of them, which I took in a flight simulator, the situation required that I initiate a pilot abort just after lift-off. This would give me a chance to experience the same reaction I would have if I had to break away from the Redstone and come straight back down in an 8,000-foot fall, plummeting like a brick. Right in the midst of this test the communications systems goofed, and Al Shepard, who was monitoring my flight as capsule communicator, could not know that I had already aborted. As far as he knew, I was doing fine, and he read his lines the way we had rehearsed them for a normal mission.

"Liberty Bell Seven," he called. "This is Cape Cap Com. Your trajectory is O.K."

"Thanks, Cape Cap Com," I called back sarcastically. "I'm just dead."

On Friday morning I moved from the Holiday Inn Motel in Cocoa Beach, where we had all been staying, and settled down in our living quarters at Hangar S, out on the Cape. I showed up early at Pad 5 for the simulated flight which would be the final practice mission in the capsule before the launch. It went fairly well, but I was kept so busy handling communications checks that I fell slightly behind in the count. All of the sequences in the countdown took place in the right order, but some of them came off a little late. Then the second of the three retro-rockets, which are programed to fire at five-second intervals, went off two and a half seconds early. So we had to check into that.

That evening I planned to be at the pad to watch the crews install the new restraining coils for the explosive hatch. But before the technicians could get started, they discovered that one of the batteries in the capsule had either gone dead or had shorted out. They were able to replace it without tearing up the capsule, so the schedule did not get set back. But we decided to rerun the simulation test on Saturday, just to make sure that the new battery was working fine and that we had not fouled up the circuits when we fooled around with the wiring.

On Saturday John and I went on our low-residue diet so that no matter which of us went on the flight we would not find ourselves embarrassed by too many natural functions. We had our booster review that day with Dr. Jack Kuettner, one of the former German scientists at Huntsville who is the Mercury test manager there. Kuettner read off a full history of the Redstone's performance since it had arrived at the Cape. Apparently the bird had been behaving well in all of its tests, for they had encountered no troubles with it. Then we had another meeting to decide if we were still agreed on all of the things we had already decided to agree on. This is one of the necessary hazards of being involved in a research program. You have to make sure all along the way that everyone who is going to push buttons or perform any other function is fully briefed on what everybody else is doing so there will be no confusion.

John Glenn ran a test of the hydrogen peroxide system which activates the control jets on the capsule. Then I climbed into the cockpit for the rerun of our last simulation flight. This time I was able to keep up with the schedule, and we did not miss a trick. In addition, all three retrorockets fired right on the button this time, five seconds apart.

Sunday was a fairly lazy day. We had a low-residue breakfast of strained orange juice, filet and poached eggs—with no coffee, tea or other stimulant. Then I went to the aeromedical area to try out some new wrist rings which had been made for my pressure suit. These were metal rings which could slide around on ball bearings and give me more freedom of movement with my wrists. That afternoon John and I went down the road to Patrick Air Force Base to begin our final physical examinations. When that was over I went back to the Cape and tried my luck at fishing. Jack Jackson, one of the doctors who was staying close to me, caught a nice yellowtail. I got nothing. I called Betty before I went to bed, and she apparently had no problems. I talked to both Scott and Mark. They had been swimming and playing baseball. Everything was normal.

I also spent a quiet day on Monday the seventeenth—the day before the launch was scheduled to go. I relaxed in our quarters, went to a short briefing in the afternoon and was in bed by five that evening. At 10:30, Bill Douglas woke me to say that the launch had been scrubbed because of bad weather.

There was nothing much for me to do the next morning but to go fishing again and try to relax. I caught one bass, but the fish did not seem to enjoy the weather any more than I did, so I threw my one fish back and went off to another meeting.

At least they had been able to decide about the scrub before they put the liquid oxygen fuel into the Redstone, which meant that the minimum delay would be only 24 hours instead of 48.

That night I watched Wyatt Earp on television and was in bed by nine o'clock. I slept like a brick for about four hours and woke up wondering what time it was and what the weather was like. Just then Bill Douglas came in and sat down on the bed. He just sat there for a few moments, but when he saw me looking at him he said simply, "Well, get up."

"How is it?" I asked.

Bill told me they had pushed the count ahead by one hour and were aiming at a 7 A.M. launch to try to beat the weather. I was sleepy, but it occurred to me that this meant we were really going to try it.

"Good," I said, and got out of bed.

The schedule allotted me 30 minutes for breakfast, another 30 minutes for a short physical, 10 minutes to fasten on the electronic sensors which would report my pulse, temperature and breathing rate back to the Control Center, 30 minutes to put on my pressure suit, 25 minutes to test the pressure in the suit, then 30 minutes to get into the van and ride out to the pad. I had saved 45 minutes by shaving and showering the night before. Apparently someone forgot to pass the word about the earlier launch time, because breakfast was not ready at 1:45 as it was supposed to be. We decided to go ahead with the physical exam first, then eat. I put on my bathrobe and went into the medical room.

There was nothing unusual about the exam except that Bill Douglas could not believe my blood pressure count.

"It can't be this low," he kept saying.

"Well," I volunteered, "I can try to boost it up a little for you."

"No," Bill answered, "but I think you ought to be just a *little* bit excited."

I will have to admit that I wasn't.

Just before breakfast I had a short session with George Ruff, our consultant headshrinker. He made me recite my feelings, and then we played some little games with words and numbers—to make sure I was completely sane, I guess. I also used this time to finish packing and to show John Glenn what things he could take down to Grand Bahama Island for me so I could change clothes when I got back from the recovery ship. I wanted to iron out all these little details so there would be nothing on my mind later.

I had breakfast with Scott Carpenter, John Glenn and Operations Director Walt Williams, the man who would monitor all the factors, including the Redstone, the spacecraft, the telemetry and the weather, and decide if and when I went. Then, at 2:55, the doctors started to glue the sensors onto my body. They had tattooed little marks on our skin so they could find the right place each time. The only thing I can remember about this period was that everybody seemed to be winking or grinning at me. They seemed to feel they had to cheer me up. I thought to myself that they were the ones who looked a little shaky, and it tickled me to realize that almost everybody in the place was nervous but me. Because of the late breakfast, we were 10 minutes behind schedule getting the sensors on. We figured I could dress a little faster than usual to make up the time. They were already working on the bird, and my arrival at the pad would have to coincide with the progress they were making on the Redstone. I had the suit on at 3:10, and after a pressure check we got into the big transfer van at 4:15. It was the same van I had ridden out to the pad with Al Shepard. It had a spare pressure suit and spare sensors in case something went wrong with mine during the 20-minute ride. We could change suits and check the pressure on the way. We had one new touch this time, however. Someone had stenciled a sign just inside the door of the van.

"Shepard and Grissom Express," it said. I got a kick out of that.

When we got to the pad, we waited inside the air-conditioned van until the word came for us to go up the gantry elevator and climb into the capsule. Deke Slayton, who was stationed in the blockhouse that morning, came into the van to give me the final weather briefing. It didn't look too good for making visual observations through the capsule window. The entire Atlantic coast from Canaveral on north would be obscured by clouds. But there would be only a four-tenths cover over Cuba, and I might get a glimpse, Deke said, of the southern tip of Florida. The local weather reports were more reassuring. People kept coming into the van to say the sky was clear overhead. It looked as if we were going to keep going.

At 5 A.M. the word came to go on up the gantry. I stepped out of the van, took a quick look at the tall, white Redstone, and headed for the elevator. Just then all the men working around the pad started to applaud. I must admit this choked me up a little. It was a darn fine feeling, as I looked down and saw them staring up at me, that I had all those people pulling for me. Bill Douglas handed me a crossword puzzle book which he'd gotten from Sam Beddingfield, the same NASA engineer who gave the crayons to Al Shepard. Sam seemed to think we needed gifts to ease the routine a little. There was a note in the book signed by Sam, who knew that my flight plan was a little less cluttered than Al's had been so I would have more time to look around.

"Gus," the note said. "Since the flight load has been reduced, we did not want you to get bored."

When the elevator reached the third level on the gantry, where the capsule sat, I walked across the platform and climbed into the cockpit. Joe Schmitt, our suit technician, began to strap me into the contour couch. I was still calm and relaxed, but as I looked around everybody was still giving me those nervous goodbye grins. The count was going nicely, and I think I could have taken a short nap just then if they had let me alone on the radio. They seemed to feel they had to keep me company and buoy up my spirits. I did not feel this was necessary, but I went along with it anyway.

The gantry crew stuck their hands inside to bid me goodbye just before they bolted on the hatch, and John Glenn slipped me a note which read, "Have a smooth apogee, Gus, and do good work. See you at GBI (Grand Bahama Island)."

I could see the troops peering at me now through the window and the periscope. I was a little concerned about all the fingerprints and noseprints on the window and I said so. Guenter Wendt, the McDonnell capsule chief, promised me they would be cleaned off before I left. One of the technicians who had overheard my complaint came on the air and said they would install windshield wipers for the next shot.

The gantry pulled back on schedule and the count was still progressing. But through the window I could see some thin clouds moving in. The weather changes very fast around the Cape, and I was a little concerned. I figured we could take off, however, if we did not get involved in any long holds. At T minus 10 minutes and 30 seconds I heard them call a five-minute hold to check the weather. This was a disappointing sign, and I wondered why they had waited so long to review that. Then, when we went into a 30-minute hold, I hit my lowest point of the day. I knew then that the weather was not going to get any better. We had to stop the procedure at some point to give the recovery ships downrange plenty of time to find me and fish me out of the water before it got dark. So I was prepared for the scrub, and it was not long in coming. They brought the gantry back, unbolted the hatch, and I climbed out and went back to the van and back to the hangar. I was getting out of my suit when Walt Williams, who had to make the decision, stopped by to explain why we had not gone. He said the weather was not good enough for the flight we had planned. There was nothing I could do but agree with him.

I was disappointed, however, after spending four hours in the couch. And I did not look forward to spending another 48 hours on the Cape. It would take that long to purge the Redstone of all its corrosive fuels, dry it out and start all over again. But I felt sure we would get it off the next time around. And we did.

The buildup was normal. I got up at 1:10, had breakfast at 1:25, let the doctors look me over again at 1:55, put the biomedical sensors on at 2:25, suited up at 2:35, climbed into the van at 3:30, and was in the spacecraft at 3:58 A.M. I was to lie there for three hours and 22 minutes before we finally lifted off.

The doctors reported that Grissom's weight just before the flight was 150 pounds and 8 ounces, his pulse 68, his respiration rate 12 breaths per minute. His neck was "normally flexible," his thyroid gland was "unremarkable." The abdomen was "soft, without tenderness." "Eye, ear, nose and mouth examination was negative." "Heart sounds were of normal quality, the rhythm regular." The psychiatrists reported: "No evidence of overt anxiety, that Astronaut Grissom explained that he was aware of the dangers of flight, but saw no gain in worrying about them. He felt somewhat tired, and was less concerned about anxiety than about being sufficiently alert to do a good job."

We had a few problems with the countdown on this try. One of the explosive bolts that held the hatch in place was misaligned, and at T minus 45 minutes they declared a hold to replace it. This took 30 minutes. Then the count was resumed and proceeded to T minus 30 minutes, where it was stopped so the technicians could turn off the pad searchlights. It was daylight by this time anyway, and the lights were causing some interference with the booster telemetry. There was another hold at T minus 15 minutes to let some clouds drift out of the way of the tracking cameras. This one lasted 41 minutes. I spent some of this time relaxing with deep breathing exercises and tensing my arms and legs to keep from getting too stiff. We finally got to the last act and I heard Deke Slayton count down to ". . . five . . . four . . . three . . . two . . . one." Then, just at lift-off, Al Shepard remembered the José Jiménez record that had helped relax him, and he gave me a parting cheer over the radio.

"Don't cry too much," Al said.

The lift-off was very smooth. I felt the booster start to vibrate and I could hear the engines start. Seconds later, the elapsed-time clock started on the instrumental panel. I punched the Time Zero Override to make sure that everything was synchronized, started the stop watch on the clock and reported over the radio that the clock had started. I could feel a low vibration at about T plus 50 seconds, but it lasted only about 20 seconds. There was nothing violent about it. It was nice and easy, just as Al had predicted. I looked for a little buffeting as I climbed to 36,000 and moved through Mach 1, the speed of sound. Al had experienced some difficulty here, his vehicle shook quite a lot and his vision was slightly blurred by the vibrations. But we had made some good fixes. We had improved the aerodynamic fairings between the capsule and the Redstone, and had put some extra padding around my head. I had no trouble at all, and I could see the instruments very clearly.

The telemetry tapes, which recorded Grissom's physical reactions every 15 seconds, showed that his pulse rate ranged from 65 to 116 beats per minute during the countdown and rose to a high of 171 when the retros fired. His breathing rate rose from 12 to a high of 32 breaths per minute during flight. His hearing was good throughout the flight, and though he had the feeling that he had to speak too fast in order to record all of his impressions, the tapes, which recorded 81 separate messages from Grissom, came out clear and distinct.

I did experience a slight tumbling sensation when the Redstone engine shut off at T plus 142 seconds and when the escape tower went 10 seconds later. There was a definite feeling of disorientation. But I knew what it was, and it did not bother me. I could hear the escape rocket fire and the bolts blow that held the tower to the capsule. And I could see the escape rocket zooming off to my right. I saw the tower climb away, and it still showed up as a long slender object against the black sky when I heard the firing of the posigrade rockets that separated the capsule from the Redstone. I could hear them bang and could definitely feel them kick. I never did see the booster, though. Neither had Al.

Now, I was on my own. Shortly after lift-off I went through a layer of cirrus clouds and broke out into the sun. The sky became blue, then a deeper blue, and then—quite suddenly and abruptly—it turned black. Al had described it as dark blue. It seemed jet black to me. There was a narrow transition band between the blue and the black—a sort of fuzzy gray area. But it was very thin, and the change from blue to black was extremely vivid. The earth itself was bright. I had a little trouble identifying land masses because of an extensive layer of clouds that hung over them. Even so, the view back down through the window was fascinating. I could make out brilliant gradations of color—the blue of the water, the white of the beaches and the brown of the land. Later on, when I was weightless and about 100 miles up—almost at the apogee of the flight— I could look down and see Cape Canaveral, sharp and clear. I could even see the buildings. This was the best reference I had for determining my position. I could pick out the Banana River and see the peninsula which runs farther south. Then I spotted the south coast of Florida. I saw what must have been West Palm Beach. I never did see Cuba. The high cirrus blotted out everything except the area from about Daytona Beach back inland to Orlando and Lakeland, to Lake Okeechobee and down to the tip of Florida. It was quite a panorama.

At one point, through the center of the window, I saw a faint star. At least I thought it was a star, and I reported that it was. It seemed about as bright as Polaris. John Glenn had bet me a steak dinner that I would see stars in the daytime, and I had bet him I would not. I knew that without atmospheric particles in space to refract the light, we *should* be able to see stars, at least theoretically. But I did not think I would be able to accommodate my eyes to the darkness fast enough to spot them. As it turned out, John lost his bet. It was Venus that I saw, and Venus is a planet. John had to pay me off, after all.

The flight itself went almost exactly according to plan. I had a really weird sensation when the capsule turned around to assume retro-fire attitude. I thought at first that I might be tumbling out of control. But I did not feel in the least bit nauseous. When I checked the instruments, I could see that everything was normal and that the maneuver was taking place just as I had experienced it on the trainer.

Just as this turnaround began, a brilliant shaft of light came flashing through the window. This was the sun. I knew it was coming, but when it started moving across my torso, from my lower left, I was afraid for a moment that it might shine directly into my eyes and blind me. Everything else in the cockpit was completely black except for this narrow shaft of light. But it moved on across my body and disappeared as the capsule finished its turnaround.

I did have some trouble with the attitude controls. They seemed sticky and sluggish to me, and the capsule did not always respond as well as I thought it should. First, the yaw control tended to overshoot a bit. And this, combined with the fact that I was trying to study things through the window at the same time, slowed me down on my program. I tried to hurry to catch up, but now the controls overshot a bit on the pitch axis. I hit the yaw rate again, but overshot once more. All of this meant that it took longer for me to work the controls than I had planned, and when my time for testing them was up I was slightly behind schedule. I wanted to fire the retros manually and at the same time use the manual controls to stay in the proper attitude. This was not critical on my flight, since I was on a ballistic path to begin with and we were just exercising the retro-rockets for practice. But it did indicate that we still had a few improvements to make with the controls. Actually, even if I had been in orbit, I could have handled the situation. It was not serious. It just wasn't perfect. This was the main reason I was up there, of course—to find the bugs in the system before we went all the way.

I was looking out of the window when I fired the retros manually, right on schedule. I could see by checking the view that a definite yaw to the right was starting up. I had planned to use the view and the horizon as a reference to hold the capsule in its proper attitude when they fired. But when I saw this yawing motion start up, I quickly switched back to instruments. You have to stay right on top of your controls when the retros fire, because they can give you a good kick in the pants and you cannot predict in which direction they may start shoving you. Here was where some extra training on the ALFA would have come in handy. It would have given me more confidence in the window as a visual reference for the controls, and I would not have felt it so necessary to go right back to the instruments that I knew best.

It was a strange sensation when the retros fired. Just before they went, I had the distinct feeling that I was moving backwards—which I was. But when they went off, and slowed me down, I definitely felt that I was going the other way. It was an illusion, of course. I had only changed speed, not direction.

Despite my problems with the controls, I was able to hold the spacecraft steady during the 22 seconds that it took for the three retros to finish their job. Then, right after the retro-package jettisoned—at T plus 6 minutes and 7 seconds—and the dead rockets fell away, I looked through the periscope and saw something floating around outside that looked just like a retro-motor. Bits and pieces of the retro-package floated past me a couple of times. It had come loose, just as it was supposed to, and had left the heatshield clean and uncluttered for re-entry.

The re-entry itself, which I knew could be a tricky period, was uneventful. But it did produce some interesting sensations. Once I saw what looked like smoke or a contrail bouncing off the heatshield as it buffeted its way through the atmosphere. I am sure that what I saw were shock waves. We were really bouncing along at this point. I was pulling quite a few Gs—they built up to 11.2. But they were no sweat. I had taken as many as 16 in the centrifuge, and this seemed easy by comparison. I could also hear a curious roar inside the capsule during this period. This was probably the noise of the blunt nose pushing its way through the atmosphere.

Both the drogue chute and the main chute broke out right on schedule. The drogue came out at T plus 9 minutes and 41 seconds, and I saw the canister fall away and watched the chute deploy. There was a mild shock when it opened, and I could also feel a pulsation inside the capsule as the air rushed through the slits in the chute and it began to breathe. Twenty-three seconds later, the drogue chute pulled the main chute out. I could see the big one first in its reefed condition. And then, when it opened up, I could see about 80 percent of it through the window. It was a very encouraging sight. There was a slight bouncing around when it dug into the air, but this was no problem. The capsule started to rotate and swing slowly under the chute as it descended. I could feel a slight jar as the landing bag dropped down to take up some of the landing shock.

I hit the water with a good bump. This came at T plus 15 minutes and 37 seconds. The capsule nosed over in the water, and the window went clear under. Almost immediately, I could hear a disconcerting gurgling noise. But I made a quick check and could see no sign of water leaking in. I thought about ejecting the reserve chute on top of the capsule to take some of the weight off and bring me back to an even keel a little faster. But I decided to see if it would right itself. And sure enough, it came around in about 20 or 30 seconds.

I felt that I was in very good shape. I had opened up the faceplate on my helmet, disconnected the oxygen hose from the helmet, unfastened the helmet from the suit, released the chest strap, the lap belt, the shoulder harness, knee straps and medical sensors. And I rolled up the neck dam of my suit, a sort of turtle-neck diaphragm made out of rubber which we tighten around the neck when we take the helmet off to keep the air inside our suit and the water out in case we get dunked during the recovery. This was the best thing I did all day.

This procedure left me connected to the capsule at only two points: the oxygen inlet hose which I still needed for cooling, and the communications wires which led into the helmet. Now I turned my attention to the hatch. I released the restraining wires at both ends and tossed them to my feet. It occurred to me that I might need the survival knife which was fastened to the door, so I removed it and stuck it into my survival kit to the left of my couch. Then I removed the cap from the detonator which would blow the hatch, and pulled out the safety pin. The detonator was now armed. But I did not touch it. I would wait to do that until the last minute, when the helicopter pilot told me he was hooked on and ready for me to come out.

I was in radio contact with "Hunt Club," the code name for the helicopters which were on their way to pick me up. The pilots seemed ready to go to work, but I asked them to stand by for three or four minutes while I made a check of all the switch positions on the instrument panel. I had been asked to do this, for we had discovered on Al's flight that some of the readings got jiggled loose while the capsule was being carried back to the carrier. I wanted to plot them accurately before we moved the capsule another foot. As soon as I had finished looking things over, I told Hunt Club that I was ready. According to the plan, the pilot was to inform me as soon as he had lifted me up a bit so that the capsule would not ship water when the hatch blew. Then I would remove my helmet, blow off the hatch and get out.

I had unhooked the oxygen inlet hose by now and was lying flat on my back and minding my own business when suddenly the hatch blew off with a dull thud. All I could see was blue sky and sea water rushing in over the sill. I made just two moves, both of them instinctive. I tossed off my helmet and then grabbed the right edge of the instrument panel and hoisted myself right through the hatch. I have never moved faster in my life. The next thing I knew I was floating high in my suit with the water up to my armpits.

Things got a little messy for the few minutes that I was in the water. First I got entangled in the line which attaches a dye-marker package to the capsule. I was afraid for a second that I would be dragged down by the line if the capsule sank. But I freed myself and figured I was still safe. I looked up then and for the first time I saw the helicopter that was moving in over the capsule. The spacecraft seemed to be sinking fast, and the pilot had all three wheels down in the water near the neck of it while the copilot stood in the door trying desperately to hook onto it. I swam over a few feet to try and help, but before I could do anything he snagged it. The top of the capsule went clear under water then. But the chopper pulled up and away and the capsule started rising gracefully out of the water. Now I thought, "These boys have really saved us after all."

I expected the same helicopter crew to drop a horse collar near me now and scoop me up. That was our plan. Instead, they pulled away and left me there. I found out later that the pilot had a red warning light on his instrument panel, telling him that he was about to burn out an engine trying to hold onto the capsule. Normally, he could have made it. But the capsule full of sea water was too heavy for him, and he had to cut it loose and let it sink. I tried to signal to him then by waving my arms. Then I tried to swim over to him. But by now there were three other choppers all hovering around trying to get close to me, and their rotor blades kicked up so much spray that it was hard for me to move. The second helicopter in line was right in front of me, and I could see two guys standing in the door with what looked like chest packs strapped around them. A third guy was taking pictures of me through a window. At this point the waves were leaping over my head, and I noticed for the first time that I was floating lower and lower in the water. I had to swim hard just to keep my head up. I thought perhaps the neck dam was leaking, so I tried to check it. Then it dawned on me that in the rush to get out before I sank I had not closed the air inlet port in the belly of my suit, where the oxygen tube fits inside the capsule. Although this hole was probably not letting much water in, it was letting air seep out, and I needed that air to help me stay afloat.

I thought to myself, "Well, you've gone through the whole flight, and now you're going to sink right here in front of all these people."

I was certain that if I did go under there would not be time for anyone to save me. I wondered if there were sharks around. Then I remembered that the chopper pilots had told me how they could drive sharks like cattle with the wash from their rotors. I also remembered the souvenirs which I had stored in the left leg of my suit, and I wished I could get rid of them. I had brought along two rolls of fifty dimes each to give to the children of friends, three one-dollar bills, some small models of the capsule and two sets of pilot's wings. These were all adding weight that I could have done without.

I wondered why the men in the door of the chopper did not try coming in after me. I remembered the time during recovery training at Pensacola when Deke Slayton got caught away from his raft with his helmet on. He was taking water and going down fast. But Wally Schirra and I, who were swimming nearby with flippers on, were able to get to him and hold him up with no sweat. I wanted Deke or someone to do the same thing for me now.

I suppose the crew did not realize how much trouble I was in. I was panting hard, and every time a wave lapped over me I took a big swallow of water. I tried to rouse them by waving my arms. But they just seemed to wave back at me. I wasn't scared now. I was angry. Then I looked to my right and saw a third helicopter coming my way and dragging a horse collar behind it across the water. In the doorway I spotted Lieutenant George Cox, the Marine pilot who had handled the recovery hook which picked up both Al Shepard and the chimp, Ham. As soon as I saw Cox, I thought, "I've got it made."

The wash from the other helicopters made it tough for Cox to move in close. I was scared again for a moment, but then, somehow, in all that confusion, Cox came in and I got hold of the sling. I realized when I threw it over my neck and arms that I had it on backwards. But I couldn't have cared less. I knew it would hold me. I hung on while they winched me up, and finally crawled into the chopper. Cox told me later that they dragged me for 15 feet along the water before I started going up. I was so exhausted I could not remember that part of it. As soon as I got into the chopper I grabbed a Mae West and started to put it on. I wanted to make certain that if anything happened to this helicopter I would not have to go through another dunking. I spent the entire trip to the carrier buckling that life jacket.

When I had been aboard the carrier for some time an officer came up and presented me with my helmet. I had left it behind in the sinking capsule, but somehow it had bobbed loose and a destroyer crew had picked it up as it floated in the water.

"For your information," the officer said, "we found it floating right next to a ten-foot shark."

This was interesting, but it was small consolation to me. We had worked so hard and had overcome so much to get Liberty Bell launched that it just seemed tragic that another glitch had robbed us of the capsule and its instruments at the very last minute. It was especially hard for me, as a professional pilot. In all of my years of flying—including combat in Korea—this was the first time that my aircraft and I had not come back together. In my entire career as a pilot, Liberty Bell was the first thing I had ever lost.

We tried for weeks afterwards to find out what had happened and how it had happened. I even crawled into capsules and tried to duplicate all of my movements, to see if I could make the same thing happen again. But it was impossible. The plunger that detonates the bolts is so far out of the way that I would have had to reach for it on purpose to hit it, and this I did not do. Even when I thrashed about with my elbows, I could not bump against it accidentally. It remained a mystery how that hatch blew. And I am afraid it always will. It was just one of those things.

Fortunately, the telemetry system worked well during the flight, and we got back enough data while I was in the air to answer the questions that I had gone out to ask. We missed the capsule, of course. It had film and tapes aboard which we would have liked to study. But since the flight itself went well, we felt that MR-4 had paid its way. We had found out a number of valuable and important things about the system, some of which confirmed Al Shepard's experiences and gave us a double check, and some of which were new:

The window proved to be the pilot's best friend. It seemed to be a great help in determining our position in space and the attitude of the spacecraft.

The sounds of activating nozzles and exploding pyrotechnics could be heard very easily inside the spacecraft, and I felt that they should offer good, dependable clues as to how the system was functioning. The flashing lights on the instrument panel would serve as a backup, and would still tell us when something was *not* taking place when it should be. But the pop, crackle and roar of the machine itself was a fine indication that its various systems were working.

The mild buffeting and vibrations which we experienced, even during re-entry, should not interfere with any of the pilot's functions on future flights. Shepard had felt this way, and so did I.

The period of weightlessness had not affected the ability of either one of us to perform our duties and manipulate the controls.

There were still a few gaps in our training which we could easily remedy—like how to make sure you don't sink when you get back.

Despite all of our headaches along the way, and an unhappy ending, the spacecraft had performed its mission. It had flown me 302.8 miles downrange, had taken me to an altitude of 118.2 miles at a speed of 5,168 miles per hour, had put me through five minutes of weightless flight and had brought me home, safe and sound. That was all that really mattered. The system itself was valid. The problems which plagued us could be fixed. And we were relieved, actually, that we had discovered them. For we were still in the development stage, when such faults and problems are to be expected. Finding them on this flight meant that we could prevent similar troubles in the future.

With our second and final suborbital mission under our belts, we were ready now for the big one.

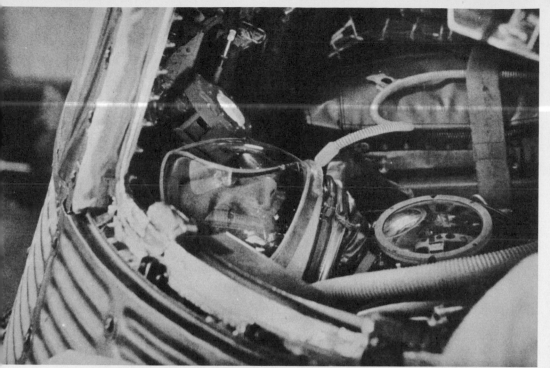

Gus Grissom lies quietly inside his capsule as the technicians get ready to lock him in for his flight. Deke Slayton [BELOW, LEFT] is in the blockhouse near the launching pad, and Al Shepard, with John Glenn behind him [BELOW, RIGHT], is at the Cap Com console in the Mercury Control Center. Shepard raises his thumb as a signal that the Astronaut is ready.

The second manned Mercury capsule, with Gus Grissom lying on his back at its base, is lifted off the pad by a Redstone booster [OPPOSITE PAGE] after one postponement due to weather. Once more spectators clustered on the nearby beaches [ABOVE]. And once more the flight went well. Grissom had some trouble with his control system, but the re-entry was uneventful and Grissom's capsule, which was named Liberty Bell 7 and had a crack painted on it in memory of the original bell, came down in the Atlantic in the prescribed area.

Then things went wrong. The hatch on the capsule popped open before it was supposed to, and, with sea water rushing in through the opening, Gus quickly jumped out into the water. He nearly drowned before the recovery helicopters finally gave up trying to save the sinking capsule and rescued Grissom in the nick of time. He was flown to the carrier, and it was a wet and dejected-looking Grissom [RIGHT] who took the telephone call from President Kennedy. The President spoke for everyone. He was relieved, he said, that Gus was safe.

THREE TO

John H. Glenn, Jr.

Malcolm Scott Carpenter

Alan B. Shepard, Jr.

MAKE READY

Four months after Gus Grissom's ballistic flight on a Redstone booster, NASA announced the team of men who would prepare for the first orbital mission. John Glenn would be the pilot; Scott Carpenter would serve as his backup pilot; Alan Shepard would act as technical advisor. In this chapter the three Astronauts describe how they prepared for the flight. John Glenn begins the chapter.

A DETAILED PLAN

John H. Glenn, Jr.

NATURALLY, it was a great moment for me when the announcement came that I would be the first American to orbit the earth. I first knew it when Robert Gilruth, the Director of Project Mercury, called the seven of us into his office at Langley and told us that I would be the pilot on MA-6 and that Scott Carpenter would be the backup pilot. After that we all stood up, and one by one the other men came over and shook my hand. We are not a bunch of back-slappers. You are always happy for the guy they pick and sorry if it's not you. Then you congratulate the pilot, and that's all there is to it.

I had hoped when the meeting was called that this would be the outcome. I had had similar feelings before, however, and they had not worked out. I had hoped all along that I would be the first man in the world to go into space. I felt that I was qualified, and I think I must have tried a thousand times since I joined the program to imagine what it would be like to sit on top of a booster, ready to go. Some disappointments were in store for me, however. First, the Russians worked faster as a nation than we did and beat us to it. Then two of my colleagues were picked to go ahead of me on the Redstone ballistic flights.

Those were rather rough days for me when I first learned that Al Shepard and then Gus Grissom would climb into their capsules and take off for space while I stayed on the ground. I guess I am a fairly dogged competitor, and getting left behind twice in a row was a little like always being a bridesmaid but never a bride.

I felt differently, of course, when it was time for Al and Gus to go. As their backup pilot, I did my best to prepare the capsule for both of them up to the last minute, and no one was happier than I was to see those two Redstones go blasting out over the Atlantic. These were the first big steppingstones in a team program, and we knew there would be many more of them to follow and plenty of work for all of us. Al took the first daring step. Gus took the next one. Now it was my turn. Looking back, I think that being the backup man was the best training I could get. I went through the drill twice and I benefited from what both Al and Gus learned for all of us. Since I had a fairly big step to take, my job was to try to push the technique quite a bit further so that the other Astronauts could learn from me.

From a technical standpoint, the orbital mission would be quite different from the ballistic flights in several respects. For one thing, we would be using the Atlas missile as a launch vehicle in order to get up the required thrust and velocity to get into orbit. The Atlas' engines have a total thrust of 360,000 pounds compared to 76,000 for the Redstone, and the Atlas would get the capsule up to a top speed of nearly 18,000 miles per hour, which is more than three times the speed of the Redstone IRBM, which had served us well on the ballistic flights. Once in orbit, the flight would also last longer—about four and a half hours if we made all three orbits. This would be a new magnitude of space flight for the U.S., and if it were successful it would pave the way for longer voyages, eventually to the moon and beyond, just as Al and Gus had paved the way for this one. It would be a prelude to our plans for the future.

We all knew that it would be a unique test in many ways. Being launched on top of an Atlas would feel like starting out twice and going through two separate launchings in a row, one after the other. First, at lift-off, all three of the Atlas' engines would be operating—the sustainer engine and the two outboard engines which would drop off after about two minutes when their job was done. This period, incidentally, would be one of the more tense moments of the day. If anything went wrong here and we had to abort and come to a fast stop, we could have a fairly rough ride. We would still be within the dense layers of the atmosphere, and the sudden shift from rapid acceleration to a fast slowdown would subject us to a considerable number of Gs. There would be a positive G factor of about 8 as we went up. Then a negative G factor of about 12 would be added to this as we slowed down. This adds up to a total change of about 20 Gs. It would be a very realistic example of the "eyeballs in and out" rides that we had taken on the centrifuge. The body would be thrown against the restraining straps, and the head would flop forward with a sudden jerking motion. Then, when the booster engines dropped off about 40 miles up, and the sustainer continued to work alone, the G forces would fall back to about one and a quarter. The Atlas would be getting lighter in weight during this period, however, because it would be burning up its load of fuel and liquid oxygen, In addition, the escape tower, which we would not need once we got above the atmosphere, would jettison when the boosters dropped off and lighten the load even more. This loss in weight would permit the Atlas to accelerate with just the sustainer engine operating, and would mean that the G forces would start building again as the load kept getting lighter until the booster would be pushing us along at about 8 Gs when the moment of truth came and we finally went into orbit. The weight of my body would actually double, redouble and redouble again in a matter of minutes during the launch. We knew that a double start like this would amount to a new sensation of flight which Al and Gus had not experienced on their Redstone missions, and part of my job would be to report back on how it felt so we would know what to expect on future flights.

Both the people on the ground and I would be extremely busy monitoring the flight during these first few minutes to make sure we were getting a good start. We would have to compare notes fast and arrive at a number of quick decisions. Al Shepard, who would be stationed in the Mercury Control Center and keeping in close touch with the tracking stations scattered around the world, will discuss this phase of the operation at greater length in the next chapter. We would also have a tremendous amount of automatic and electronic equipment to rely on. The Abort Sensing and Implementation System (ASIS), which Deke Slayton described earlier in the book, would be keeping track of the booster itself and would trigger the escape rocket immediately and pull the capsule free if it detected any imminent catastrophe in the booster. I could activate the escape tower myself by moving the abort handle with my left hand if I did not like the way things were going and there was not enough time to talk things over or wait and see if the automatic system had spotted the same trouble. The monitors in the Control Center or at the Bermuda tracking station could also initiate an abort if they detected deviations in the trajectory or in the performance of the booster. In other words, I would have a lot of help. In the months preceding the launch all of us rehearsed our parts in the mission over and over to make sure we all understood what the procedures would be at every moment and that nothing was left to chance.

Much time was also spent working up a detailed flight plan which plotted each event that would take place during the mission and each duty and activity that I would try to perform along the way. This plan was the result of many meetings between various NASA engineers and scientists and myself. We wanted to make the most of every moment that I would be off the ground, and so we went over the flight plan many times, moving events from one period to another like pieces in a puzzle. The important thing was to wind up with a schedule which would give us everything we needed to know about the operation of the capsule, about my own reactions, and about the environment of space, without loading it so full that we jeopardized any part of the mission. The main priority was a safe and sound journey. During the launch, for example, I was scheduled to report many things to the ground by radio, including a description of my own status plus the vibrations that we knew I would be running into. I would also give a detailed run-down on such operational workings of the capsule as cabin pressure, oxygen quantity, fuel levels, battery power and any other observations I wanted to make. All of this data would be displayed on huge lighted panels in the Control Center so that the personnel there could compare it, minute by minute, with what the situation *should* be at each stage of the flight and make decisions or recommendations concerning it. They would also have a tremendous amount of information flooding in on them from other sources—by telemetry from the booster itself, and from the radar and photographic tracking stations all over the world which would give them an accurate picture of the trajectory the booster was taking. Even though I might feel inside the capsule that I was doing fine, the people in the Control Center would have the big picture in front of them and could exercise their option at any point during launch to bring me back. I would not be on my own just yet.

The event we would all be waiting for would be the exact moment when I got into orbit. This would occur about five minutes after the launch, when I would be closer to the Bermuda tracking station, actually, than to the Control Center at Canaveral. It would come when we reached the point of "insertion," and hitting this keyhole in the sky with the capsule would be like trying to fire a bullet through a small knothole at a thousand yards without grazing the wood. All the factors of the flight —the azimuth, the angle of insertion and the velocity of the booster— would have to be right on the button or we would not get into a proper orbit. There was little tolerance for errors. In order to conserve fuel, we wanted to achieve an elliptical rather than a circular orbit. We wanted to go high enough, however, to get above the atmosphere in order to eliminate the drag that would slow us down. At the same time, we did not want to go too high or we would come close to the dangerous Van Allen belts of radiation which lie several hundred miles above the earth and could expose us to more cosmic particles than we had been able to protect ourselves against in the capsule. Taking all these problems into account, we planned for an apogee, or high point, of about 145 miles and a perigee, or low point, of about 85 miles. Our success would depend in part on the speed of the Atlas. If its velocity were too low, for example, we would not get into an orbit at all but would simply re-enter the atmosphere and come back down into the Atlantic again somewhere in the vicinity of the Canary Islands. We would be picked up here by ships if this happened, but it would be a pretty crazy suborbital flight.

Another factor in our success would be the direction and angle of our trajectory. This would depend, in turn, on the guidance of the Atlas.

The system for controlling the trajectory is a complex one. During the powered phase of flight, for example, while both the outboard boosters and the sustainer engine were working together, an electronic programer inside the Atlas would guide it along the prescribed path. The angle of flight would be predetermined and fed into the programer in the form of electronic commands placed on tape. These commands would keep it on course if any forces—like strong winds, for example—tried to deflect it. As soon as the Atlas left the ground, radar tracking equipment would lock onto us and start feeding data on the flight into huge computers located at the Goddard Space Flight Center near Washington. The computers would analyze this information and compare it in a matter of seconds with data they had already stored up which showed how the flight *should* be proceeding. They would have a complete picture of how well the booster was aiming the capsule at the keyhole. Then, at 2 minutes and 34 seconds after launch, when the booster engines and the escape tower would jettison, the Atlas would continue under the power of its sustainer engine. At this point—which we call BECO for "Booster Engine Cutoff"—the guidance controls would shift from the Atlas to the ground. The computers there would now take over completely and start sending up whatever commands and corrections were necessary to keep the Atlas on its proper course.

The sustainer engine in the Atlas can be gimbaled to keep it on the proper trajectory. That is, the engine can swivel back and forth to keep the booster headed in a certain direction—the way you balance a broomstick on your hand by moving your wrist back and forth underneath the stick. If the Atlas strays off course, the computer radios up commands which cause the engines to swivel just enough to straighten it out. If all goes well and the Atlas finds the keyhole, the computer commands the sustainer engine to shut itself off—we call this SECO, for "Sustainer Engine Cutoff"—and then the explosive bolts which hold the capsule and the Atlas together are blown automatically and the posigrade rockets fire to separate the capsule from the Atlas. At this point, which would come about 100 miles up and five minutes and five seconds after the launch, we would be on our own. We might or might not get the all clear to proceed into orbit, however. The people on the ground would still be making one final check of the computers and the data in front of them to make sure that we were moving at the correct speed of 17,500 miles per hour to sustain an orbit, and I would continue to fill them in on the operation of the capsule. The automatic control system would already be releasing hydrogen peroxide fuel through the nozzles to turn

the capsule around a full 180 degrees and place it in its orbital attitude—
with the blunt base pointed forward and slightly upward, instead of
straight down as it had been during the launch. I might feel a slight tumbling sensation here as we turned around and started to ride backwards,
and the flight plan called for me to report on this. A light should have
turned green on my instrument panel by now to indicate that the Cap
Sep or "Capsule Separation" event had occurred on schedule. The periscope should have extended automatically. I would then make a complete
check of all of the electrical systems to make sure that they were working and report on their status to the Control Center. Meanwhile, the computers on the ground would be rapidly analyzing the radar data to determine whether the conditions were right for insertion into orbit. Finally,
at a point between six and six and a half minutes after launch, I would get
the word through Al Shepard in Mercury Control as to whether or not
I had a complete "Go" and was actually headed for a proper orbit. This
would be an exciting moment, but the flight plan called for some vital
work to be done at this point, so there would be no time to spare. Al
would also give me the exact times at which the retro-rockets would have
to be fired to start bringing the capsule home at the end of one, two or
three orbits. The computers would already have calculated this data,
based on our speed and angle of insertion, and even though I would not
be using this information for some time, it was so critical for the success
of the mission that the plan called for me to confirm the retro-fire information and check it against the charts I had with me before I got busy
with any other matters. It would be a little like riding a commuter train
on a very bad night, I guess. You would want to make sure before you
settled down that you were ready to get off at your station, where people
were waiting for you.

As we continued to fill in the flight plan, we estimated that I would
approach the west coast of Africa some 16 minutes after launch. I would
want to mark this crossing through the periscope, record the exact moment by pushing a telemetry key and note my altitude as I went over. I
would carry out a similar operation as I reached other specific checkpoints
along the way in order to confirm the precise timing of the orbit, and
keep track of exactly where I was each moment of the way in case I
suddenly had to come down. About 28 minutes after launch, I would be
in contact with the tracking station at Kano, Nigeria, and would start
to prepare the cabin for darkness, which would come, amazingly enough,
in another 15 minutes. The flight plan included a prescribed routine for

this event, which included making sure that my flashlight was readily accessible and that the cabin lights were dimmed and covered with red filters so I could quickly grow accustomed to the darkness and would have no trouble reading the instruments and maps. Two minutes after I finished this work, I would have been on my way for half an hour and would have to make the first of a series of reports that the people on the ground would be expecting from me every 30 minutes during the flight. The exact timing of the reports would depend on how clear my radio contact happened to be at the time. I also had a prescribed checklist for this event, and I rehearsed it many times during our training to make sure that I would be able to do it almost automatically when the time came without leaving anything out. It was a long list, and included ticking off the status of everything from fuel and oxygen consumption to the positions of the switches, the temperature, pressure and relative humidity in the cabin and the volt and ampere readings on all the batteries.

Approximately four minutes after this first full report was made, I would start running down another checklist, this time to prepare for firing the retro-rockets in case an emergency occurred in one of the systems to make us want to come down before we reached the end of the first orbit. As Al Shepard will explain later in this chapter, we had a number of contingency areas plotted along the entire route where we might need to land in the event we could not quite make it to the end of the line where the major recovery forces were waiting. We had to be prepared to land in each of these areas as they came up, just in case. At about 33 minutes after launch, for example, I would be in communication with the tracking station on Zanzibar, the large island off the east coast of Africa. The next recovery area—which we call Contingency Area IC—was a long way off, in the Indian Ocean, near the west coast of Australia. In order to be prepared to land there, however, I would have to start various activities right after I passed over Zanzibar, or I would not be ready to fire the retro-rockets on time. The loose equipment would have to be stowed, for example, so it would not float around and get in my way during an emergency re-entry. I would have to check the attitude control system to make certain that I could use it to get into the proper attitude before I fired the retro-rockets. This would include warming up the hydrogen peroxide thrusters so they would go to work without delay when I activated the controls. I would have to confirm the position of a number of switches so that everything would take place in the proper sequence. The restraint harness would need adjusting so that

I would be prepared for the shock of a possible landing. I would double-check the attitude of the capsule by comparing the readings on the instrument against the actual horizon outside, which I could see by looking through the window and the periscope. Finally, I would verify the time on the various clocks once more, so I could be sure that the rockets would fire right on the dot.

All of this work was scheduled so that it would take only about a minute. By the time it was over, according to the flight plan, I would have left Zanzibar behind, and it would be time for me to observe my first sunset of the day. We estimated that this event would start 43 minutes after launch. A few minutes later I would check our capability to align the attitude of the capsule by lining it up against the night horizon. We were fairly certain that the moon would help us here. Eight minutes after I checked the horizon, I would have crossed the vast Indian Ocean and would be approaching the west coast of Australia. Here, I would start the retro-fire checklist again, this time to be ready to come down, if necessary, in Contingency Area ID, which was set up near the Fiji Islands in the Pacific. Five minutes after that I would make my second 30-minute report of the flight—to the Woomera tracking station in Australia—try my hand at tracking the stars for a while to see if this could be done, have my first meal, and then start through the retro-fire and re-entry checklists once more. This time I would be preparing to land in the normal recovery area which had been designated in case we decided to end the flight after the first orbit. I would still be over the Pacific at this point, and the recovery area would be several thousand miles away in the Atlantic as we talked about it. But the retro-rockets have to be fired exactly 2,990 miles from the spot where you intend to land. This meant making the final "Go-No Go" decision while I was still some 600 miles off the California coast. As we calculated this on the flight plan, the decision would come one hour and 27 minutes after I had taken off. Ten minutes after that, if everything was still normal and we had decided to continue, I would be on my way along the second orbit and would be in direct contact once more with Al Shepard at Cape Canaveral, right back where we started.

The flight plan for the second and third orbits was worked out with equal precision. I would no sooner be contacting Al Shepard, for example, than it would be time to start thinking about making an emergency landing in Contingency Area 2A, which is near the west coast of Africa. Then, during the next 37 minutes, I planned to carry out various tests to determine my reactions to weightlessness, make a daylight horizon check over the Canary Islands, rehearse the retro procedure for Contingency Area 2B, which was off the east coast of Africa, observe the earth below while I carried out some yawing maneuvers for approximately four minutes, prepare the cabin for darkness again, make my fourth 30-minute report, observe the second sunset of the day, get ready for emergency re-entry in Contingency Area 2C, which was off the west coast of Australia, and look down as I passed over the tracking ship stationed in the Indian Ocean to see if I could spot some bright flares which the ship was scheduled to send up by balloon during the night.

I also planned, during the second orbit, to have a second meal. This would not be because I was hungry, but simply to make another test of man's ability to consume food in a state of weightlessness. We also set aside some time during which I would test my vision, take my blood pressure once more, and do some more navigating by the stars as we came across the dark Pacific. Here, I intended to pitch the spacecraft up approximately 34 degrees from its orbital attitude, choose a star which fell more or less along the center line of the window, then slowly pitch the capsule down again while I tried to hold the star in the same position with respect to the window. This maneuver would help me determine whether we would be able to count on navigating by the stars on future flights.

On the third orbit, I planned to make detailed observations of the weather down below, watch my third sunset, carry out further tests of my vision, blood pressure and reactions to weightlessness, then start warming up the thrusters for the last time as we approached Hawaii. Here I would go through a fixed-routine check of the retro-fire procedure before we actually fired the rockets that would bring me home. The retro-rockets themselves would slow the capsule down by about 500 feet per second. This would be just enough loss in speed to decrease the centrifugal force that had held the spacecraft in orbit and permit gravity to start bringing it back into the atmosphere. This would slow it down even more, until finally it came down in the recovery area that we had predicted for it. We knew, from our previous tests and calculations, that the capsule would slow down from 17,500 miles per hour to 270 miles per hour in about 10 minutes on this final leg and that the heat pulse surrounding the capsule would build up to about 9,500 degrees Fahrenheit, which is very close to the temperature of the sun. It would be a warm homecoming.

It seemed incredible to me as we scheduled these events on the flight plan that I would soon be doing all of this. It seemed fantastic that during my transcontinental speed record run in a jet plane I had a top speed of 1,100 miles per hour and would now be covering the same distance almost eighteen times as fast. I would fly across the United States—from San Diego to Savannah on the second orbit—in just eight minutes. The capsule would be moving so rapidly from sunlight into darkness and back into sunlight again that an entire day would go by in about 40 minutes and a whole night would pass in another 40 minutes. The sun and the moon and stars would set eighteen times as fast from my orbital position as they do here on earth. Thinking about flying that fast was like my trying to visualize a million dollars. There was just nothing in my experience to which to relate it. But there it was, all written down in the flight plan. It still seemed incredible.

I suppose that I had been looking forward to this trip for so long that as the time grew near, I could scarcely believe it. The same reaction is probably true for anyone who plans something big and worth while over a long period of time and then suddenly sees it coming to pass. I know I have felt the same way, for example, about my family. Most men probably do. Having a family was something that I looked forward to long before I was married. It has always been a vital part of my life, and I have some strong ideals on the subject. Now, whenever I stop to realize that I have a wonderful family and that some of my ideas have somehow worked out, it is hard to believe it.

As I have explained earlier in the book, however, there would be nothing spooky or supernatural about this flight. I was fully aware that there might be some dangers involved. These are bound to exist in any venture like this that is so new and untried. I knew that a person could get hurt in this business. At no time, however, was I physically afraid. I have always had the idea that people who are afraid of what will happen to them personally whenever they think about attempting something new will seldom do or dare very much—or take the risks which are necessary to bring on progress. In addition to this, I have a religious faith which adds to my feeling of confidence in a situation like this. I do not happen to believe in a fire-engine kind of religion which encourages a man to call on his Maker only when he needs help. I do not think that God orders every detail of my life and that He will see that I always come through, no matter what I do. I have always thought that my mother has a good slant on this. She has always had a strong Christian faith, but she feels that our relationship to

God is a fifty-fifty proposition. God placed us on this earth, she believes, with certain abilities and talents. How well we use them is completely up to us. My own family and I share these beliefs, and the confidence this has given us in the future has helped absolve our worries.

As far as I was concerned, this mission would probably come closer to using all of the skills, talents and instincts that I possess than anything I had ever attempted. It would mean being something of an explorer, pioneer, pilot and scientific guinea pig all at once. I knew that some of our scientists had criticized Project Mercury as a waste of time and effort. They had claimed that instruments and little black boxes could bring back as much information as a man could on this kind of flight, so why send a man? I definitely did not agree. I did not happen to believe that black boxes, no matter how clever they were, could substitute for the curiosity of an explorer, the maneuvering ability of a trained pilot or the judgment of a human observer. These were points that I hoped to prove. To be sure, the first orbital mission would only be a scouting trip. As we built more powerful boosters we could look forward to the time when we could fly deeper into space and carry out all kinds of valuable scientific missions. My job at the moment was to give this particular capsule and its systems a thorough workout, to find out what if any physical effects this kind of venture into space might have on a man and to bring back recommendations which might help our engineers start planning the design of future space vehicles and systems. We had much to learn about the limitations and capabilities of man in space, and this would be a unique opportunity to put some of our questions to the test. I planned, for example, to put myself through regular periods of physical exercise and run up my pulse rate to determine the real effects of weightlessness on my system. We added a heavy bungee cord to the equipment in the capsule, and I planned to pull it as far as it would go at certain periods during the flight to see what effect a known amount of exercise would have on my heart. I planned some maneuvering of the capsule to determine whether my senses were affected in any way. I would also be taped with several kinds of sensors, just as Al Shepard and Gus Grissom had been, so that the specialists in aeromedicine who would be monitoring my physical reactions during the mission could compare these with similar readings made before and after the flight. This data would help determine how a man with known characteristics and human tolerances can cope with such forces as high Gs, weightlessness, isolation and heat which I would undergo. This was one reason why I tried to keep in good physical shape before the flight. I wanted to make sure that I was ready.

I suppose the thing that sustained me most as we prepared for the flight was that I had complete confidence in the mission. First of all, I had confidence in the people who were backing it up. This would be a "we-type" operation. Thousands of scientists, industry specialists and technicians stationed at the Cape, at the tracking stations around the world and in the recovery fleet at sea would be working together as one huge team to make the trip possible. I had worked with them for so long that I knew they would do their best. They would include teams from General Dynamics who had built the Atlas and would help launch it, crews from McDonnell and NASA who had checked out the capsule and would get it ready on the morning of the launch, and, of course, the six other Astronauts with whom I had worked so closely for three years and who would man key positions during the flight.

I had confidence in the detailed plans that we had made and rehearsed so many times, and I had complete faith in the hardware that we would use. I had kept in close touch with my capsule from the time it was being assembled in St. Louis, and I stayed with it through most of the many stringent tests that the engineers gave it once it arrived at Hangar S at the Cape. I got to know the capsule inside out during the long weeks of training that piled up during the delays, and both Scott Carpenter and I spent as many as ten hours each in one day working inside it, along with the McDonnell engineers and the NASA personnel, while we put it through its paces. As Scott will explain in greater detail later in this chapter, we ran off simulated missions so many times before the day of the launch that once we got going it would seem like just another dry run. That, as all of us have tried to point out in this book, is the purpose of a good training program—to practice until the real thing seems almost easy by comparison.

I also had confidence in myself. I knew that I was ready to go, and this, too, was important. There was a lot of teamwork and "togetherness" involved in the mission. At the same time, however, there was a certain feeling of aloneness about the flight. There would be only one man in the capsule, and once we got into orbit there would be no one around to do all the things that had to be done except the pilot. Along this line, some friends of mine who like to joke asked me on several occasions if I was not worried about the capsule's number. It so happened that Friendship

7—a name that my son and daughter, Dave and Lyn, picked out of a list that we made up together—was the thirteenth Mercury capsule to come from the factory. It was carried on all the lists, therefore, as Capsule 13. Since I am as completely unsuperstitious as any man I know, the only answer I could give to my friends was that so far as I was concerned, any number they wanted to give my capsule—including 131313—could bode nothing but good.

A CLOSE COLLABORATION

Malcolm Scott Carpenter

JOHN GLENN and I knew each other so well by the time we teamed up together on MA-6 that we really never had to stop and discuss how we would manage the partnership, or who would do what. As it turned out, I think I may have sat in the capsule a little longer than John did during the weeks of preparation, because I tried to take care of most of the routine checks on booster and capsule systems while John concentrated on practicing orbital missions in the simulator.

We thought alike on almost everything we did, and I was amazed at how many times John's solution to some problem turned out to be identical with what mine had been. We also saw eye to eye on the scope of the mission itself. We both viewed the flight from the beginning not just as a matter of sending a test pilot up to report back on how his machine operated, but as a magnificent opportunity to prove what man could do in space and to bring back information that the weather people and the astronomers had been crying for. Our management did not always agree with us. They felt that the primary purpose of this first manned orbital flight was to prove out the system and that many of the scientific observations should be left for a later flight. Perhaps it was just as well. As it turned out, John had his hands fairly full just flying the capsule for two entire orbits, and he had to leave undone a number of experiments which he had intended carrying out.

We had a long time, of course, to get ready. The spacecraft John was to fly arrived at the Cape near the end of August 1961. Originally, he was scheduled to fly the mission that December. Then it was postponed until the following January, and it finally got off the ground on February 20, 1962, after a number of delays caused by weather and technical problems. We used every minute of this time to prepare ourselves and the equipment for the flight. We spent a total of 166 working days, for example, running checks on the capsule and making changes in it and repairs right up to launch time. Some of the changes were brought about by things that we learned from the two-orbit flight of Enos, the chimp, which preceded our first manned orbital flight. All told, we made 225 different changes in the capsule itself. I will discuss some of these in more detail a little later, but they ranged all the way from installing a new auto-pilot to reinforcing the fuse holders on the instrument panel because of mechanical failures we had experienced on previous tests.

John and I participated in all of the systems checkouts and we reviewed all of the changes in design, so that nothing was done without the pilot's knowing about it in detail. John gave me leeway from the beginning, however, to make decisions for him whenever I could, and I was happy to take quite a bit of the detail work off his shoulders. Even so, he was a hard man to keep up with. John is a tremendous driver. And he was so completely dedicated to his job that he did not seem to need the rest and relaxation that I did. He could think and eat Project Mercury twenty-four hours a day for weeks at a time without letting anything else intrude on his time. Frankly, it wore me out sometimes just trying to keep up with the pace that he set for himself. One of the duties which I tried to perform, as a matter of fact, was to persuade him to let up a little and take a break whenever he could. We are both tremendously fond of music, and this common interest helped. I had a guitar and a ukelele with me at the Cape, and occasionally after work we would do a little harmonizing together. Or we would go out to dinner somewhere where there was a little band and ask them to play something we both liked. I remember one evening when we sat for a couple of hours listening to a Polynesian combination. John liked the lilting sound of one song in particular, and he asked them to play it for us several times before we left. It was "Beyond the Reef." I jotted down the chords on a paper napkin and added this to my rather slender repertoire. Another song we liked was "Moonlight in Vermont." John could really sink his tenor into that one. As for me, I was particularly fond of "Yellowbird," and John liked to join me in singing this, too.

But evenings like this were rare. For one thing, as we came closer to launch time, it was just too much trouble for John to try to go anywhere in public. Reporters and photographers would close in around him, even when he went to church on Sunday mornings. And people kept coming up to ask for autographs. John is one Astronaut who does not mind this sort of thing in the least; in fact, he feels it is part of his job, and he goes out of his way to be thoughtful to anyone who comes up to him. But still, he could hardly get much relaxation in this atmosphere. And he was also a little worried that he might pick up a stray bug if he mingled in too many crowds. So he stayed close to his quarters in Hangar S on the Cape, and some of the rest of us tried to make it pleasant for him by joining him on the Cape for dinner. Our Air Force dietician, Miss Bea Finkelstein, took over the kitchen duties near flight time and did a marvelous job of varying our low-residue diet as much as possible. One night, when we did not have to be too careful, she prepared sweet and sour pork with rice and vegetables and served it with a little hot *sake*. Our secretary, Miss Nancy Lowe, drove into Cocoa Beach to bring back some chopsticks and a Chinese lantern to make it more festive.

John and I did a good deal of our own cooking, however—like a couple of bachelors whose wives are out of town. Near the end, when we were working constantly from morning to night, we often got through the day with nothing more for lunch than a couple of cans of Metrecal and per- haps a bowl of soup. This was to save time, not for reducing. John was staying in shape by getting a lot of exercise. Also, we got a daily "Care" package from Henri Landiwirth, the very thoughtful manager of the motel where we normally stayed in Cocoa Beach. Henri knew that we both liked shrimp—with extra hot sauce—so he sent some out to us almost every day by whoever was coming. He included some of the black bread that John is very fond of, and some dried-up potato skins. That was a joke. Henri's chef had a habit of serving baked potatoes without the skins and John had kidded Henri so unmercifully for skimping on potato skins that Henri thought up this gesture to get even with him.

So the time passed not too unpleasantly. I knew that some people think John must have gone stir-crazy during the long delays. He even got a couple of get-well cards in his mail which read, "Sorry to hear about your confinement." But John was of such a single purpose that he turned each delay into an advantage and used it to hone the edge of his readiness a little finer. We were originally scheduled to launch in December of 1961. We could have tried it, but some of us thought this would be press- ing things unnecessarily, and NASA decided to take the pressure off and aim instead for early 1962. It was completely typical of John Glenn that even with this extra time he took only four days off to go home for the holidays—two for Christmas and another two for New Year's—which meant that he had to travel Christmas night and New Year's night in order to get back to work.

Because this was the first time that a team of Astronauts had ever prepared for an orbital flight, we developed our own set of procedures. We were very methodical about it. I started out by going through all of the diagrams of the wiring—we have already described the seven miles of that—and came up with some new ideas for arranging some of the sequences of events which would help streamline the flight plan. As I went through each system, circuit by circuit, I also started compiling long lists of questions on how the various systems operated and how they might differ from the training capsules that we were familiar with. This was a rather monumental job. The capsule was still so new and completely untried as far as manned orbital flight was concerned that my list of questions grew very long. I sent various sections of it to the appropriate sources of information—the McDonnell capsule people, the autopilot people, the environmental control people, the communications people, etc. They got together detailed reports which answered the questions, and then we had a series of meetings in the aeromedical area on the Cape to discuss these answers. I think that everyone understood the systems much better after these meetings were over, and that we all benefited. As a result of them, in fact, we made quite a few changes—some small, some large—in several of the systems.

We had had some trouble on previous flights, for example, with the inverters, the little devices that convert the regular D.C. power of the batteries in the capsule to A.C. power in order to run the fans, the cabin lights and the automatic stabilization and control system. The inverters had tended to overheat once in a while, and the problem was to find out—while you were in flight—which one of the two sets of inverters was causing trouble. It was a complicated process the way it was set up originally; you had to run down quite a long checklist of switches and fuses. This was not exactly foolproof, for it was conceivable that you could upset the one inverter that was working fine while you were looking for the one that wasn't. We hit on a simple way to find out. We would insert a couple of extra little wires into the capsule which would turn a light on when the inverter they were attached to was working. We were already using these wires as temporary gadgets anyway when we ran preflight checks on the system. We simply decided to make the wires a permanent fixture. One light was labeled "Fans," the other was labeled "ASCS" (for the automatic stabilization and control system). We also made a simple change in the cooling system so that the inverters would not get so hot in the first place. We installed a duct leading to the inverters from the fan which normally keeps the cabin cool. The fan would suck hot air away from the inverter and help keep it cool, too. In addition, we put some small cold-plates around the inverters and routed water through them, just as we do to cool the pressure suit. This meant adding a separate temperature control knob to control the flow of water so it would not get too hot and make excessive steam. We were concerned here that too much steam could result in water condensing around the vent where the steam was supposed to leave the capsule and freeze it shut. This would have plugged up the entire cooling system and given us more trouble with the inverters than we had had to start with. We thought we made a good change here, and, as it turned out, we apparently did. The inverters did overheat during John's flight and we had to make even further changes in them, but they were no real problem.

The attitude control system, which was to bother John on his flight, checked out very well during the weeks of tests and simulations. We had a few minor problems during this period with other items. One of the two fans in the pressure-suit circuit was faulty and we replaced it early in the game. The satellite clock installed in the capsule jammed one day during a test, and we replaced it, too. John also discovered that two of the microphone wires in his helmet were reversed, and the helmet had to be repaired. These were the kinds of problems that are bound to crop up in a program like Mercury where all your equipment is so new, and this was the reason why we were not in a great hurry to rush this flight. We wanted to run enough tests to make sure that we had caught all of the problems on the ground.

While this work was going on, John and I kept busy taking refresher courses in the various skills that an Astronaut has to have to make his mission worth while. We went to Pensacola, Florida, to become familiar with night vision at the U.S. Naval Aviation Medical Center there and ride around in a revolving room to become accustomed to the feeling of being disoriented. We spent two very productive days at the Moorehead Planetarium in Chapel Hill, North Carolina, studying star patterns and constellations with the astronomers, and we reviewed this work on our own night after night, with a transparent globe. This had the sun and stars marked on it, and we held it up against the dark sky to help find the stars we would use to orient ourselves by when we were in orbit on the dark side of the earth.

John also did some extra training in recovery procedures. No one had ever figured out just what happened to Gus Grissom's capsule to make the hatch blow prematurely, a mishap which resulted in losing the capsule and almost losing Gus. We were all determined to plug that loophole, and Al Shepard, who was our team expert in recovery procedures, helped work out a new technique. It involved dropping a team of two SCUBA divers into the water from the first recovery helicopter to reach the scene. The divers would immediately fasten a buoyant bag under the capsule which would keep it afloat even if filled with water. As soon as the bag was in position, one of the divers would crawl up on the capsule and help John try to climb out through the narrow neck on top. If that procedure did not work, they would blow the side hatch and get him out that way. John and I also went out into the Atlantic two or three times to work with the emergency raft that we carry inside the capsule, just to get more familiar with it in case John had to use it. Because John is a thorough man about everything, he went over the survival equipment he would carry in the capsule with a fine-tooth comb. This included a bag of shark repellent, dye markers, a first-aid kit, a desalting kit, distress signals, a signal mirror, a jackknife, a pair of sunglasses, emergency rations, a container of water, some waterproof matches, medical injectors—including some morphine in case of injury—a length of nylon cord and a signal whistle to attract attention at night. We also added a new gadget to the suit as a result of Gus's flight. This was a tiny life vest which weighed less than a pound and could be folded up into a package not much bigger than a man's hand. It was fastened onto the chest of John's suit and he could have inflated it and stuck his arms through it in less than 10 seconds to keep himself afloat. All of these items had to be thoroughly tested, of course, not only for their reliability in an emergency but also for their ability to withstand all of the stresses and strains of space flight, including shock, vibration, acceleration and the intense heat and cold they might be subjected to before they would be used. As it turned out, of course, John did not need any of this equipment. But we had to worry about it during the training phase and become familiar with each piece of it, just in case.

In the meantime, I went on drawing up new lists of questions, making pages and pages of notes of things to do—which I checked off as each item was taken care of—helping draw up the work schedules and keeping a close eye on the status of the special equipment John would carry. One of my notes to myself was to make sure that someone remembered to clean off the periscope and window so that John would have a clear view for navigation. There were some fairly technical questions, too. I asked the technicians, for example, whether it was necessary for the pilot to abort the mission immediately if for some reason the escape tower failed to separate from the capsule.

The answer to this question was "No," and I passed it on to John. We had both discussed this one night and were not sure of the details ourselves, since the matter had never come up. I also ran down someone who could confirm for us the exact distance from the point where the retro-rockets fired in a circular orbit to the capsule's point of impact in the recovery area. Normally, this would concern only the people on the ground who would send up the radioed commands to initiate retro-firing. But if something went wrong with communications and John had to compute this critical sequence for himself, he would need to have the exact data at his fingertips. The answer to this question was 2,990 miles.

Getting back to the troublesome inverters again, we were not positive whether they should be turned off manually if they overheated. The answer here was, "Yes, when they get to 185 degrees." I was curious about whether some little rubber hoses in the capsule had been inspected by the engineers to make sure they had not rotted and were in good shape. I was told I could assure John they were in good condition. I made notes to check on the new radio code which John would be using—the dot-and-dash which he would resort to if we lost voice communication—and to see that a new lightweight life raft which was being provided at the last minute had been thoroughly tested and qualified before the flight. It was made of new material and even had ballast buckets mounted on the bottom to help steady it on rough seas. I discussed most of these items with John as they came up, so he would know in his own mind the things that had been done and would not have to worry about them himself. I also kept a list of things to remind John about, like making sure he reserved some time in his flight plan for checking the pressure on his helmet visor seal bottle before he opened it. Test pilots—and Astronauts—live by checklists. They have so many complicated things to do and remember that sometimes the little, obvious things do not get done simply because they are not written down in the proper sequence in the flight plan. It is a meticulous approach, and sometimes it verges on nit-picking. But it is a good way to stay alive and fly another day.

Another example of the detailed, meticulous approach which we have all taken towards space flight was the long list of mission rules which the operations people drew up to govern every conceivable event and activity, both normal and abnormal, that could occur before and during the flight. These were rules that were binding on all of us, and they were written down so that everyone who had anything to do with the flight knew exactly what they were. They ran all the way from general guidelines to obscure little points that might have cropped up only once in a thousand times.

One general guideline, for example, was that whenever possible the Astronaut would try out all of the systems which were vital to the mission and as many of the backups as he could on the first orbit. This was so we would know exactly what we could rely on in case we got into trouble. John's flight plan reflected this rule; it specified a number of tests which he would carry out on the control system, the various communications channels, etc.

Another mission rule involved communications. It stated that if the Astronaut lost the ability to transmit by voice to the Control Center during the launch, we would abort the mission and re-enter immediately. The operations people felt that the success of the mission depended so heavily on the sound of the Astronaut's voice and on the subtle undertones of the reports he would be making about himself and the systems he was using that they did not want to let him go any further if they could not hear his voice. The rule went on to add that if he had achieved orbit *before* we lost voice communications, we would let him go for at least the duration of that orbit.

Both John and I objected to this rule. The operations staff—in particular, Flight Director Chris Kraft, whom Al Shepard will introduce more fully a little later—wanted to abort any launch on which we lost voice contact, regardless of when it happened. Chris's point was that the only alternative would be to communicate with the ground through dot-and-dash code, which he did not feel was adequate for reporting on such complex matters as the malfunction of a complicated system in flight. If we could not actually hear the Astronaut's voice coming over, Chris believed it was wiser to initiate an abort. John and I discussed this problem during various meetings. We were convinced that it was an unnecessary precaution. Then, one day when I was inside the procedures trainer making a simulated flight, the simulator crew pulled a fast one on me. They shut off all voice communications. I think they thought I would goof and prove Chris's point. But I went to work with the code key and managed to send him all of the information he needed. I also told him by code that I wanted to go ahead and complete the mission. It was only a practice run, so Chris let me try it. We made out fine.

They pulled the same thing on John one day when he was in the trainer. He had no warning, either, but he switched over to the key immediately and started banging out all of the data they needed in code. So we won our point. We would not abort the launch if the Astronaut lost his voice transmitter. We did go on to develop the code more fully, however, until we were satisfied we could use it to describe anything that was happening in the capsule as well as we could describe it with our voice. Chris did hold out two provisos, though. He would let the mission proceed without voice after insertion *if* all of the other systems were "Go," and *if* he was completely satisfied from the telemetry information he was receiving that the pilot's aeromedical condition was 100 percent sound. This was the main problem that Chris was worried about, I think—that without being able to use his voice the Astronaut could not describe his own condition as fully as Chris would like.

As we have mentioned earlier in the book, safety—particularly safety for the Astronaut—was a paramount rule in Project Mercury. Perhaps we Astronauts were willing to take a few extra chances in order to get moving; that was the way we were built. But our bosses had their eyes on a bigger picture. They were not willing to *let* us take chances which they felt might jeopardize the mission or discredit the program. All of life, of course, is a compromise. And what we were doing was life at its fullest.

We also had to memorize quite a few mission rules involving technical procedures and equipment in the capsule which would be important once the mission got under way. We all agreed, for example, that if one of the two small fans inside the pressure-suit circuit failed after launch, we would continue with the flight. If both fans failed, however, we would turn on the emergency oxygen flow and re-enter at the end of that orbit. The reason for this order was that we carried only enough oxygen for one orbit at this increased flow rate. We could manage with one fan, however, since the second one was primarily a backup anyway. We also had rules covering the use of the faceplate on the helmet. If it should fail to reseal, for example, after we had opened it to eat, and we were having trouble closing it again, we should make plans to re-enter at the end of that orbit. This was because the suit could no longer be pressurized with a leak in the helmet, and we were now dependent on cabin pressure alone to keep us alive. The rules also stipulated that if the cabin pressure started to fall at the same time that we were having suit problems, we should switch on the emergency oxygen flow at once and prepare to land immediately in one of the contingency areas instead of waiting for the normal recovery area at the end of that orbit.

Throughout all of this work, of course, I had to make sure that I would be ready to fly the mission myself if John should become ill and NASA decided to switch pilots. I went through all of the training that he did. The only difference was that John had priority on the simulator, since it was highly unlikely that he would not make the flight and simulator time was extremely important to the success of the mission. The two of us spent a total of about 100 hours in simulated flight during our training for MA-6. Of this, John spent about 60.

We also had a lot of homework to do. We both read and reread all of the manuals to make sure that we had all the procedures and sequences straight in our minds. We took turns sitting in the trainer for hours at a time, getting used to the feel of it and trying out various combinations of control motions and switching. We concentrated on the retro-fire maneuver, since this was the most critical. We would need this skill desperately if the autopilot was not working during retro-fire and we also lost the automatic mechanism which fired the rockets. In this case, we would have to fire the retros ourselves and control the capsule manually so it would maintain the correct attitude. This is not an easy trick, and it took a lot of practice.

We rehearsed the countdown over and over, practiced the steps we would take in the event of an abort, and went through one SEDR after another. These were the long and extremely involved series of tests which we gave each system in the capsule to make sure it was functioning perfectly. This abbreviation for "Systems Engineering Department Report" is pronounced "cedar." These tests included SEDR Number 76, which is a complete checkout of the electrical power system; SEDR 69, which runs through all of the automatic control machinery; SEDR 83, which is the final checkout of the environmental control system; SEDR 65, which gives you a good look at the instrumentation; and SEDR 93, which is a check on the weight and balance of the capsule.

Near the end of the series, just before you take the capsule out of the hangar to mate it with the Atlas, you put it through SEDR 77. This is a crucial test of all of the systems under every conceivable condition. It takes four full 24-hour days to run off—if all goes well—and during part of it John and I put in two and a half days continuously without a break. It was quite a bit like rehearsing for the opening night of a difficult play, I imagine. In the case of SEDR 77, the script is so precise and exhaustive

that it fills two big 8½ x 11-inch books, three inches thick. There are
three or four dozen other people on stage—McDonnell personnel and
NASA crews. Everyone is in his appointed spot, crawling over the capsule or sitting in the checkout trailers or in the white room inside Hangar
S. They all have headsets on and have turned the book open to the proper
place; and everyone keeps one eye on the telemetry to see how the capsule responds. Then you start out from page one and go right through.
You have to follow the chronology of events in unrelenting order to make
sure you don't miss a single step and that you put the capsule through the
most thorough and realistic paces anyone can devise. This is where you
find out whether you can put the show on the road. You are suited up in
your pressure suit, just as you would be for a regular mission; and the
only time you get to take it easy is if something goes wrong with one of
the systems and you climb out of the capsule long enough to let the technician take over to make repairs. In one period during the test, I put the
suit on at 2 A.M., kept it on for six hours, then took it off for a couple of
hours while the technicians did some work. I put it back on for a few
more hours, slept in it for about four hours when the test did not happen
to involve me, then went back to work again. All told, I was encased in
the suit for 24 out of 36 hours. It is not a comfortable kind of activity, but
you push yourself hard during a SEDR 77 and you don't let the discom-
fort deter you.

One of the most important manuals in our library is a publication called
SEDR 109. This is a pilot's handbook on how to use the capsule which is
put out by the McDonnell people who build it. SEDR 109 is a little like
the set of instructions you get with any mechanical gadget, except that
in our case it pertained specifically to Capsule 13, the one that John would
fly. The book filled a small, handy looseleaf book which was indexed
along the edge of the pages so you could find any section of it in a hurry,
and it was up to date on all of the changes that had been made in the capsule or in the procedural data that we were supposed to know when we
used it. Basically, SEDR 109 is a long series of checklists which tell you
exactly what the Astronaut should do, in time sequence, from the moment
he gets into the capsule. One of the first checklists, for example, has 17
items on it, starting out with Item 1 which tells you to connect the communications and biomedical leads that connect you to the telemetry and
voice transmitters in the capsule. Next, you run a check on the intercom
and the biomedical leads to make sure they are working. Then you connect the oxygen inlet to the suit. A little later—Item 5—you turn the suit

fan switch to "On." Diagrams of all the switches are printed in the book so you can refresh your memory as you go along. Another checklist in 3EDR 109 has to do with your activities at about T minus 90 minutes in the countdown. This is primarily an SOP for inspecting the interior of the cockpit before you go. You start out with Item 1 by making sure that the attitude control handle is centered and locked. This is just like putting your gears into neutral before you start your car. Next—Item 2—you make sure the abort handle is "Inboard" and "*locked*," which means putting this particular gear into neutral. You check to see that the survival kit is firmly secured. You unlock the shoulder harness—that's Item 4. On Item 24, which is about one quarter of the way through this particular list, you make sure the guard on the retro-fire switch is installed so the retro-rockets cannot be fired accidentally. On the last item in this list, number 85, you check all battery switches to make sure they are in the correct position.

SEDR 109 is also full of little warnings to make sure that you do not forget how the machine works and make some simple but foolish mistake. One of these warnings, for example, reads: "The cabin pressure must be at 5.0 (plus or minus 1) psi before opening helmet faceplate." This is to prevent you from losing the pressure in your suit if your cabin pressure is not adequate. Another reads: "The lap belt, chest strap and leg straps may be loosened to provide greater comfort, but should not be unfastened." Another warning reminds you to make sure that the periscope retracts "completely" when it comes inside the capsule just before re-entry. If it does not, the warning explains, the periscope door will not close tight behind it, and the cabin will overheat during re-entry. The slightly opened door would let in some of the tremendous heat which would be building up outside the capsule as you began the re-entry. If it does not retract, the periscope can be cranked manually.

While John and I were busy doing all this with his capsule, Deke Slayton was nearby starting through the same detailed procedures with his. He was several weeks behind us, of course, for his capsule did not arrive at the hangar until we had almost completed that phase of the work on Capsule 13. Deke was so wrapped up in his work and so eager to go that one of the saddest things I have ever had to do in my life was to take over his assignment several weeks later when a board of doctors finally ruled that he could not make the MA-7 flight. We had all known about the flutter in Deke's heart, but we all believed—and Deke had proved it to our own satisfaction—that it was a minor aberration which would have no effect on his role as a pilot. Deke had always performed as well as any of us during our strenuous training, and we were convinced this was an insignificant detail. But once more the top management was being cautious and Deke suffered a deep disappointment. I was happy to plunge back into work and prepare for my own flight, but I was sorry that my assignment was tinged by the great disappointment that everyone, including me, felt for Deke.

John and I had a few setbacks of our own, of course. I learned a lot about stick-to-it-iveness from him, however. One of our main problems was weather. We were in the midst of winter, and we had to wait until some lucky day when all of the various combinations of waves and wind and overcast—not only in our own launching area but in the recovery area and to a certain extent in the contingency areas scattered around the whole world—were in our favor. We also had some bothersome technical problems. Some fuel leaked through to an insulating bulkhead in the Atlas, and we had to wait nearly a week while the technicians removed a rocket engine in order to make room, then crawled up inside the Atlas on a temporary scaffolding to remove the bulkhead. About a week before that happened, we ran into some trouble with John's pressure suit. It was using up too much oxygen during one of the simulations. Mine had worked fine on a similar test. The trouble was finally traced to a leak in the gloves on John's suit. The gloves had never leaked before, so this was annoying. But the technicians fixed the problem and we got over that hurdle.

There were other problems, some of them minor and some of them time-consuming. In general, however, despite all of the fixes we made, the capsule itself was in fairly good shape all along. Everyone along the line knew that this was to be the first manned orbital capsule, and when we got it it was the cleanest, most nearly perfect capsule we had ever had. Most of the problems and delays arose when we tried to improve it.

As a result of the slips in our schedule, John was able to take time to satisfy himself about a number of other things that had concerned him. A plug in the center of the heatshield, for example, had fallen out on Enos' capsule. Everyone was fairly certain that this had happened on impact and that it would have had no effect during re-entry, when we would need every bit of that heatshield, including the plug. We used the extra time we had to have X-rays made of the shield on Capsule 13 and were satisfied that it was good and tight. We also double-checked the hydrogen peroxide system to make sure that it had no serious leaks. A leak in the control fuel system had delayed Enos' flight, and we wanted to be sure that a thing like this would not happen in the midst of John's mission.

It was a busy, even grueling time. We often finished the day tired and a little haggard, and I would try a little harder to make John get away from it for an hour or so. He had plenty to do. His mail was piling up by this time, and since John takes a sincere interest in anyone who expresses any interest in the program, he stayed up late several nights during the delays trying to make a dent in the pile of letters that was beginning to flood the office.

Finally, everything came out even. The Atlas people fixed the problems that had cropped up as our bird sat there on the pad. The capsule was clicking like a clock every time we tested it. The weather began to get better after a long, bad streak. And John was in real good shape. I think probably we were all more ready and confident of success than we would have been if we had gone earlier. At any rate, at about midnight on the morning of February 20, I left Hangar S, where John was still sleeping, and went out to Pad 14, about four miles away, where the searchlights were playing over the Atlas and where the launch crews were quietly running through the last few hours of the long countdown. I took the elevator up to the eleventh deck, stowed inside the capsule the little ditty bag full of instruments and gadgets which John would be using during the flight, and gave Capsule 13 a final check. There was a special checklist for this, too. Item 1: Knife—STOWED. Item 2: Liquid waste container—SECURED. Item 3: Six battery switches—OFF. Item 4: Launch control switch—OFF. Item 5: Food and water—SECURED. Item 6: Abort handle—INBOARD AND LOCKED.

I called back to Hangar S to report that the countdown was proceeding very smoothly and that the capsule was ready. All we were waiting for now was for the local weather to clear up and for the doctors to get John ready and send him out.

A RANGE AROUND THE WORLD

Alan B. Shepard, Jr.

WE ALL DECIDED, as we passed from the ballistic flight phase into the orbital phase of Project Mercury, that the two-man team we had had before —consisting of a pilot and a backup—was not quite adequate for the complex missions coming up. Every once in a while something got overlooked that a third person, with no specific pilot role to pin him down, could have picked up and taken care of. This was my function on John's flight. I had been through this kind of mill before. In both cases, we flew the first missions of their kind, so I knew the kind of problems that John might run into. He and Scott Carpenter would both be fairly well preoccupied with the details of the capsule itself. My job was to check through the other phases of the flight and keep myself free to go up any alley that seemed to present problems. I was sort of a combined trouble shooter and coordinator, and I spent a good deal of my time going from committee to committee and from place to place making sure that everyone knew the details of what everyone else was doing. I checked with John almost every day, and both Gus and I tried to help by advising him on what our reactions would be—or had been—under similar circumstances. The three of us could talk pilot talk together, and we did. "How do you *feel* about this solution? How long do you think it would take to get used to that new procedure? Have we gone the right way on the flight plan or should we revise it again?" Both Gus and I took some kidding from our friends on the Cape for being the elder statesmen of the project. But the fact that we had both been down this road before helped us pave the way for the work that John had to do, and everyone was most helpful. Later on, Gus performed the same role for Scott when he made his flight in Aurora 7 and acted as his Cap Com at the Cape.

John has already pointed out some of the technical differences between a ballistic flight and an orbital flight. The basic differences were that he got to ride over a much longer course and had the privilege of being the first man to ride the Atlas. The technicians had done a superb job of preparing this booster. It had been designed as a weapons system, and I would say that qualifying it to carry a man safely into space ranks as one of the major technological feats of our time.

Another new achievement that is of an equal magnitude technically is the world-wide tracking range which was constructed especially for Project Mercury. This is the series of sixteen major tracking and communications stations—counting Cape Canaveral—which was scattered around the world along the path of the orbit and is held together by one of the most fabulous communications networks anyone has ever devised. Part of my responsibility in the early days of the program was to put my two cents' worth of advice in as the engineers planned the range. And though I got fairly well immersed in other things when the time came for my mission on MR-3, I got to know quite a bit about it. I had another chance to study it when the time came for John's flight. As the technical advisor on the team, I spent quite a bit of time checking into the range that would guide John around.

The first station that an orbiting capsule passes over after it leaves the Cape is Bermuda. Our technicians decided to launch in this northeasterly direction for several reasons. One was that they wanted to take advantage of the earth's rotation. The eastward component imparted by the earth's spin would give the capsule extra momentum as it went into orbit. This angle would also permit the capsule to return over the U.S. at the end of each orbit so that we could maintain firm communications with it during this critical period from our Control Center at Cape Canaveral. By starting out on a line from Canaveral through Bermuda, we also committed ourselves to an orbital path that would lie entirely within temperate waters all the way around the world; it wound back and forth across the equator, with a maximum deviation in latitude of 30 degrees. This would come in handy in case the Astronaut had to land somewhere en route and spend a night at sea before we could find him. The path also took advantage of existing radar and communications facilities in Australia and Hawaii and the southern United States. And it would take the capsule over friendly land masses all the way—which was a political factor of some importance, since we wanted to demonstrate our peaceful and scientific intentions from the beginning and not let anyone get the idea that we had the slightest interest in high-altitude espionage.

There were a couple of large holes in this range—the wide gaps stretching across the Atlantic and Indian Oceans where the distance between land-based stations meant that we would lose communications with the capsule for a brief period. NASA filled both of these gaps, however, by outfitting two freighters with special tracking and communications equipment and stationing these ships, along with NASA personnel, in these two areas before a shoot. When the chain was completed, it ran from Canaveral to Bermuda; to the Atlantic Ocean ship *Rose Knot;* to the Canary Islands off the west coast of Africa; to the Indian Ocean ship *Coastal Sentry;* to Muchea, a small town on the west coast of Australia; to Woomera, the headquarters of the Australian government's own missile range; to Canton Island, a small coral atoll which belongs to the U.S. and is situated about halfway between Australia and Hawaii; to Kauai Island in Hawaii; to Point Arguello, California, a windy spot on top of a 2,000-foot peak overlooking the Pacific near the big missile complex at Vandenberg Air Force Base; to Guaymas, Mexico, on the shore of the Gulf of Mexico; to our old nuclear testing grounds at White Sands, New Mexico; to Corpus Christi, Texas; to Eglin Air Force Base in southwestern Florida; and finally on to Canaveral again.

We set up two stations on the west coast—Guaymas and Point Arguello —because we wanted to be well covered during the critical period when the capsule would be preparing for re-entry and would require a perfect synchronization of clocks and schedules to fire the retros on time. Guaymas was a fine position from which to monitor the first orbit; Arguello was the primary port of re-entry for the second and third orbits, which swung farther north. White Sands, Corpus Christi and the radar at Eglin would simply provide us with extra tracking information as the capsule crossed the U.S. for the third and last time and help us predict the exact impact point in the recovery area where the spacecraft would land. We fed this last-minute data to the recovery fleet so the planes and helicopters would know where to look.

Though we used existing facilities wherever we could to save time, some of the stations had to be set up from scratch. Each of them required its own power plant so we would not be at the mercy of local commercial power failures. The radar equipment had to be installed with absolute precision so we could pinpoint the position of the capsule with a maximum error in the angle of only six seconds. At Kano, Nigeria—which plugged the big gap across Africa between the Canary Islands and Zanzibar—most of the local natives who helped set up the towers and antennae and steel sheds that housed the delicate equipment did not even know what a missile was, much less that a man was about to ride over their heads in a space capsule.

Except for brief periods of silence during the third orbit, where the orbital path dips quite a distance south of Kano and Zanzibar and arches north of Woomera and Canton Island, the capsule would be in almost continuous contact with the network, and no major event could take place without our knowing about it immediately—including an emergency re-entry that we had not planned for. We were not too worried about these small third-orbit gaps. We assumed that if the spacecraft had made it this far and was in good enough shape to start out on the third orbit in the first place, it would probably stay in good shape until it could land. The world did not happen to be set up as an ideal tracking range for us, and we had to compromise somewhere and take a few calculated risks. They were fairly well calculated, however. In this program, any risk that we took was closely figured.

Six of the stations—Bermuda, Muchea, Kauai, Arguello and Guaymas, in addition to Canaveral—were designed as major command posts and were equipped not only to track the spacecraft but to send radio commands up to it, if necessary, to affect the timing of its retro-firing and re-entry. In addition, Bermuda was also set up to act as a backup Control Center in the event communications broke down between the Cape and the spacecraft during the critical period of BECO, SECO and insertion. If this happened, Bermuda would take over and make the decisions. It was equipped with its own set of computers to accomplish this task.

One amazing thing about this system was that although some of the sites were located on opposite sides of the world, we were able to trade information back and forth instantaneously. Since every second counted when a critical function like firing the retro-rockets was at stake, this was extremely important. It was made possible by the fact that the stations were woven together by a tremendously intricate communications system. This included high-frequency (HF) and ultra-high-frequency (UHF) radio channels which kept the capsule in almost constant voice communication with one station or another. In addition to conversing with the Astronaut and picking up his standard 30-minute report, the stations were also pinpointing his exact position by radar and picking up the stream of telemetry that the capsule was sending down at the rate of over 100 separate bits of data every second.

All of this information came flooding into the computers at the Goddard Space Flight Center in Greenbelt, Maryland. Each station transmitted this to Goddard over a wide variety of channels—by radio, microwave relay, leased telephone lines and cable—and used them all at once if it had them available. This way, even if some of the channels happened to get blocked by atmospheric interference or a disrupted circuit, the information was sure to get through somehow. Both Bermuda and the Canary Island station, for example, were tied in with Goddard and the Control Center by hard line cable as well as by wireless. This was the old backup principle again. The data from Nigeria and Zanzibar was radioed to London, where it was sent by cable to New York and thence to Goddard. The data from the two Australian stations was sent by cable direct to Vancouver, British Columbia, where it was put on leased lines to Goddard. It was also radioed as far as Hawaii, where it was put on the cable line that runs from there to Oakland, California, and transferred there to another leased telephone circuit. (One of the factors which we took into consideration when we picked a launch date, incidentally, was whether the atmospheric conditions on that day were likely to interfere with our communications. We tried to predict the refraction and considered the cables at the bottom of the sea as well as the clouds overhead.) The total distance between all of the tracking stations is 25,000 miles; but they are linked together by a web of wiring—including teletype circuits, telephone channels, data-transmission wires and backups for all of these—that is 140,900 miles long. Wherever two locations are connected by cable, we use a pair of cables, not just one. The computers at Goddard are connected with the Control Center, which is some 900 miles away, by eight pairs of leased lines, or 16 separate circuits, five or six of which might be in actual use at any moment. The others are on standby. We take no chances on not knowing what is going on.

There are a number of things that we have to know, instantaneously, on any orbital flight. For one thing, we must know exactly where the capsule is at all times so we can continue to plot its distance from the recovery area and update the retro times as the flight progresses. The radar tracking stations provide this data. First, it goes automatically to the computers at Goddard. The information comes in in terms of the elevation, bearing and distance at which the radar dishes spotted the capsule. These readings do not mean a thing to the computers, however, unless they can coordinate them with the exact time at which the readings were made. That is why we had to have the network of hard lines and cables—to give us what we called "real time," or instantaneous time checks on all of the data. Once the computers have added all this up, they pass the word on to us immediately at the Cape in the form of electronic messages. Then they do something even more remarkable. They prepare a series of predictions as to exactly when the capsule will appear over other tracking stations along the route and transmit this data, which we call "acquisition messages," to each station in the form of readable, teletyped messages. The computers do this all by themselves, having been previously programed by man's ingenuity. All this was rather uncanny, and despite our success at building boosters and capsules that work and training men to fly them, it seems obvious that we could never have ushered in the space age if the electronics industry had not come up with machines which can think fast and do away with a lot of the guesswork. One modern scientific breakthrough has made another one possible.

We could not have done it without telemetry, either. This is the method of gathering data at great distances and transmitting it, bit by bit, as little blips of radio energy. The telemetry started streaming out from Friendship 7 long before it went into orbit and it kept flooding into the computers until after the spacecraft landed in the ocean. Some of these facts have to do with the Astronaut himself, and they come from the sensors which the doctors anchored onto his body before he suited up to pick up his pulse, blood pressure, body temperature and breathing rate. Approximately a third of the bits give us a steady flow of information about the life support system and how well it is functioning—how much oxygen is left in the cylindrical tanks, for example, what the temperature is inside the suit, or how much carbon dioxide has built up inside the system. The rest of the data give us accurate readings on the performance of the spacecraft—its interior and exterior temperatures, the cabin pressure, amounts of fuel left, and the attitude that the capsule is holding—read off in terms of its angles of yaw, pitch and roll. This data is transmitted automatically from the capsule in a precise sequence so that the computers can keep it straight. We had to be sure, as all of this information came flooding in in such a hurry, that we did not get mixed up. We did not want to see the figure of 60 appear on the telemetry chart, for example, and get the erroneous and frightening idea that it referred to the Astronaut's body temperature when it was really a measurement of the temperature of the oxygen as it entered the inlet valve of the suit. We never did get mixed up on this sort of thing, however. Thanks to an ingenious system of electronic sorters and coders, the recorders kept track of where each bit of data had come from, when it had arrived and what it referred to. From time to time we could pause during a flight to ask the recorders a question. The machines would quickly reach back into their memory drums, answer the question, then go on about their business. They are built in such a way that these human interruptions don't rattle them a bit. They can accept as many as 32 questions in a row, one on top of the other, answer each one, then go back and pick up where they had left off without dropping a single stitch.

Even though the electronic machines were clever, we did not let them run the show. In most cases, they simply gave us the raw material for our own decisions. We were engaged, after all, in a "man-in-space" program. We had a man in that capsule whose brain was arriving at some useful conclusions, too. We had some good men on the ground, both in the Control Center and scattered around at the tracking stations, who were trained to pitch in and help. In the future, as a matter of fact, when we start probing deeper into space and get out towards the moon, the Astronaut in the driver's seat will have even more to say about his flight than he has had a chance to on these early tests of the system. In order to make a lunar landing, for example, we are going to have to let the crew be fairly self-contained. They will make quite a few of their own decisions about the velocity of their vehicle, the best way to control their trajectory and the proper moment to fire the retros, slow down, change course, glide into the moon's orbit, attempt a landing or take-off. The further you get from earth, the more valuable the input of man himself is going to be. These crews may take small computers along with them to help them with their calculus. But they will have to rely to a great degree on their own training and judgment.

The seven of us have had useful practice making fast decisions, even though we have not all taken flights. This has been especially true since the start of the orbital phase of Mercury. We had an Astronaut on duty at each of the major tracking stations, not only for John Glenn's mission —and later for Scott Carpenter's—but also for MA-5, the two-orbit flight by the chimpanzee Enos which preceded John's mission, and for MA-4, the empty capsule which preceded Enos and went around for one orbit. The main purpose of this was to let the Astronauts act as Cap Coms or Capsule Communicators at each station, and do most of the talking with the spacecraft as it passed over their area. There was not very much stimulating conversation, I'm afraid, as either the empty capsule or Enos went past. But we played it straight, and every man on each crew went down all the checklists and acted out his part just as if there had been a man aboard. There was a good reason for this. MA-4 and MA-5 were dress rehearsals for the day when MA-6 went up with John aboard, and we wanted to make a thorough test of communications between the capsule and the ground stations. As we went through each drill, tape recorders aboard the capsule broadcast our own voices back to earth so we could test the reception. We also tested it the other way around by talking back to the capsules and letting our voices be recorded in space as each spacecraft passed over our area. We played these back when we recovered the spacecraft to make sure that John would be able to understand us and that we would be able to hear him clearly when it really counted. The tapes came out fine.

During Enos' flight, I acted as Cap Com at the Bermuda tracking station. There are so few automobiles on Bermuda that most of us used motor-bicycles to get back and forth to the station. It was a rather ironic situation. Here we were, setting up a space flight that would carry a capsule around the world at almost 18,000 miles per hour, and we were obliged by law to hold our bikes down to 20 miles per hour. In preparation for each orbital flight, we tested out the range several times and ran off simulated missions to make sure that the communications were in order and that the crews knew the SOP. Each station manager had special tapes locked up in his briefcase which simulated his own station's participation in the flight. Occasionally the tapes had unexpected glitches planted in them on purpose, just to see what the rest of the range would do in a crisis and how long it would take the crews to get on the ball and figure a way out of the trouble. Sometimes the malfunction was due to a simulated power failure; sometimes it was a faulty signal. By running through these realistic rehearsals over and over we had the range in good shape when it came time for John's mission.

The range performed very well, for example, during Enos' flight. The MA-5 capsule developed several malfunctions and it had to be brought home rather quickly at the end of the second orbit. The attitude control system began to get sticky; an inverter which changes power from D.C. to A.C. for some of the equipment overheated, and the temperatures in both the cabin and in Enos' sealed compartment went above normal. The tracking network discovered these problems as soon as they cropped up. The Indian Ocean ship noticed the temperature rise; Wally Schirra confirmed it as the capsule passed over Muchea and alerted the Control Center. With less than a minute to go before the capsule got beyond the deadline for retro-firing, the Control Center transmitted the retro countdown to the tracking station at Point Arguello, where Gordon Cooper was on duty, and a NASA technician triggered the command to the spacecraft which started the retro sequence. We had a hairy moment or two on this flight when our voice communications between Canaveral and Arguello faltered as we got close to the moment of climax. But the trouble was repaired in the nick of time, and Enos got down safe and sound.

Wally Schirra was the second Astronaut to visit Muchea. Scott Carpenter had been there for the flight of MA-4. They both had a good deal of work to do, but from the stories they brought back about Australian hospitality I think they probably enjoyed themselves as much as they had at any time in Project Mercury. The Australians have been most friendly to the Mercury crews, and very cooperative about helping us in any way that they can. Scott and Wally were kept busy with simulations while they waited for the flights and we tested out the range. This kept them up until the wee hours of the morning by Australian time. Scott, who is a Colorado boy and particularly fond of horseback riding, would sleep for a few hours, then saddle up a horse which some Australian friends had loaned him and go chasing kangaroos. He caught one, too. Wally spent a good deal of his free time surfboarding, which is big in Australia, and skindiving. The Australians let him try out a new board they had devised and also took him out in a boat to frolic with some friendly seals. The nicest touch, though, came on the morning when Enos was being launched. As Wally took his Cap Com position in the operations room, he noticed that someone had thoughtfully and quietly placed a banana at each console. It was typical, understated Australian humor. The hosts just wanted to honor the American "Astronaut" of the day, who would be feeding himself on banana-flavored pellets. Wally, who would honestly have preferred being in the capsule himself to monitoring a chimp, enjoyed the joke.

The tracking stations were mainly outposts. Their function was to keep track of the capsule's position, debrief the Astronaut and keep him informed of his progress as he passed over, relay all information that they picked up to the Control Center at Canaveral and act as Mercury Control's agent in the event of an emergency. The Control Center was headquarters. Here was where all of the data wound up; here is where the decisions were made.

The room itself is about the size of an average grade-school gymnasium. It is lined with three rows of consoles which face a huge animated map of the world that runs across one wall. The map pinpoints the location of Cape Canaveral and each of the tracking stations, including the two ships at sea. Each station is circled by a ring some 12 inches in diameter which represents the normal line-of-sight range of its electronic contact with the spacecraft.

Like almost everything else in the room, the rings are color-coded so we can keep track of the status of the range just by glancing quickly at the map. If a tracking station is in a "Go" condition and is ready for us to proceed, the ring lights up green. If it is having some minor, temporary trouble, the ring is orange. If it is having serious trouble and is not ready to function, the ring is red. At the bottom of the map are several sets of large symbols, one for each station. When a particular station is on a "No Go" status, the appropriate symbol lights up to indicate which of the several facilities there is on the blink—radar, computer, telemetry, etc. The mission is held up until all of the stations are ready and most of the lights are green. We do not want to start the ball going until all of the players are in position to keep it rolling. Once the spacecraft goes into orbit, the stations pass it on from one to the other, like a baton in a relay race. Their radar devices "acquire" it first, then lock onto it and follow it until it leaves their area and heads for the next station. So long as a particular station has contact with the capsule, the green ring around its position on the map blinks on and off to let us know in the Control Center. The ring stops blinking as soon as the station loses contact.

The display also includes a small replica of the capsule which moves across the map on wires to show us exactly where the real capsule is in relation to the earth. A red light moves along the same course, ahead of the capsule, to point out the impact area where the capsule would land if the retro-rockets were fired at that particular moment. The movement of both the symbolic capsule and the red light that precedes it is governed by the computers at Goddard. There is no little man back there making guesses and pulling strings. The computers control the entire display.

I suppose that all of these rather dramatic props should make for quite a bit of drama in this room on the day of a flight. We do have our moments of suspense and worry—we had some on John's flight. But actually, we have watched the rings flash and the props move so many times during simulations that when the real thing comes along we are much too engrossed by other things to be awed. There are very few men in the room. I happened to be one of them on the day that John Glenn made his orbital flight. I sat at the Cap Com spot on the first row of consoles, with a headset around my ears, a microphone anchored in front of my lips and a series of dials and colored lights mounted on a black panel in front of me.

At the console to my right was Don Arabian, the Capsule Systems monitor. As his title indicates, he would keep score on the electrical power, fuel supply and control systems in the capsule and would be the first to notice if anything vital had gone wrong on board it. On a big display in front of his console—to the right of the huge map—he could watch, as we all could, a continuing display of data that was coming back by telemetry which showed us, among other things, the amount of fuel remaining in the spacecraft for both the manual and automatic control systems, the number of watt-hours of power remaining in the batteries and the amount of voltage that the batteries were putting out to activate the various electrical circuits. To my left was Frank Samonski, the Capsule Environment monitor, who kept his eye on the telemetry data that told us what the pressure was inside the spacecraft, how the air-conditioning system was working, how warm it was inside the capsule, how much oxygen was left in both the regular tank and the emergency or backup supply. To his left was Dr. Stanley White, the Project Mercury flight surgeon, who would keep tabs on John's physiological condition. Posted on the big board in front of Stan—to the left of the map—was the latest data on John's heart rate and respiration rate, his body temperature and the pressure and temperature inside his suit. On the console directly in front of him, Dr. White could watch John's heartbeat as it darted across a green-colored oscilloscope screen in the form of a squiggly electronic line. With this device, in fact, the doctor could study John's heart action at almost the same instant it was beating 150 miles up as the spacecraft passed overhead. He would be the first to notice any change in John's condition that might concern us.

In the back of the room, overlooking everything, was the associate director of NASA's Manned Spacecraft Center, Walter Williams. Walt is an old hand at flight testing. He had been in charge of the X-15 program, among others. Now, doubling in brass in the Control Center as our operations director, he had over-all responsibility for the launch. Walt had sweated out all of the delays more than any of us, perhaps, because he was the man who had to decide whether we tried to go on any particular morning or not. For several days before each launch date, he studied the weather reports and the status of the booster and the capsule, and discussed the pros and cons with all of us. Then, on the morning of the launch, when all of the pieces finally looked as if they might be fitting together, he took his place at the operations console and put on his headset. Walt could see everything that we could see on the big boards and the map. He also had some high-priced help. On one side of him sat Rear

Admiral John Chew, of the U.S. Navy, who was in command of the task force of 24 ships and more than 60 aircraft that made up the recovery force. On the other side of him was Major General Leighton Davis, of the U.S. Air Force, who commanded the Canaveral missile-launching complex. Through his headset and mike, Walt kept in touch with the blockhouse at Pad 14, three miles from the Control Center, where the launch crews were making last-minute checks of the Atlas and loading John into the spacecraft. The actual launching would be triggered at the blockhouse, for this was where the instruments were located that kept track of the Atlas' pulse down to the last second. But the blockhouse was under Walt's control. Obviously, they could not push the button until Walt was satisfied that the weather, the tracking range, the recovery fleet and John himself were ready.

In the far right-hand corner of the Control Center was one of the most important and complicated consoles in the room, presided over by the flight dynamics officer, Tec Roberts. Tec kept his eye on huge moving sheets of paper and a series of pens that marked off the launch path and velocity of the Atlas as it took off, and watched for any deviations in the trajectory. As he watched the pens trace their lines—the computers were moving them—Tec could begin to predict in his own mind whether the Atlas was achieving the desired thrust and staying on the proper course to head for that keyhole that John has described. He was the first man in the room to know whether we had a chance of an orbit or not. If the velocity had built up too fast, it was Tec's job to recommend we consider the abort procedure. If the Atlas had deviated from its path, it was Tec who would have to decide whether it could be brought back in time to thread the needle or if it would miss the insertion point no matter how strong a guidance signal we sent to it.

Directly behind me, between my console and Walt Williams' slot, was one of the most important positions of all—the flight director's console. This was manned, as always, by Christopher Columbus Kraft, Jr.—who is better known in the trade as Chris Kraft. Like his boss, Walt Williams, Chris doubles in jobs. Normally, he is assistant chief of flight operations for the Manned Spacecraft Center. During all launches, however, he sits at the flight director's console. This is really the hot seat once we get going. There is not much time for committee meetings when the seconds are ticking by, so Chris has sole responsibility for making most of the fast decisions. He has the authority to abort the mission—a decision that would have to be made almost before two men could even mention it.

And once we are in orbit he makes the continuing decisions about whether we keep going. Chris is also an old hand at flight-test programs, and he knows the ways of test pilots so well that he can usually stay ahead of them. On the morning of a launch, Chris is searching for bits and pieces of decisions from everyone in the room. He is the focal point now for most of the events and the flow of information. He can hear several voices at once in his headset. He can see on the big board how things are going without wasting time talking. But there are always moments when a little consultation helps. There are so many things to watch and keep in mind. Scott Carpenter has already discussed the mission rules that we had to follow. One of the most important of these involves the pressurization of the cabin. As Wally Schirra has explained earlier, the pressure suit and the pressurized cabin serve as backups for each other. If one goes out, the other will keep the man alive—so long as it continues to function, that is. On the morning of a launch, however, both of these systems have to be functioning perfectly or we don't go. This happens to be one of the most important mission rules. The environmental control system which keeps the pilot alive must be working 100 percent before the launch, with no discrepancies. We would not purposely launch an orbital mission with one major backup out of commission any more than you would start driving across a strange desert without a full tank of gas, some extra fuel stashed away, a good radiator and an extra supply of water. Then, during the first four and a half minutes of launch, before we reach the insertion point and the "Go" or "No Go" decision as to orbit, Chris has to start thinking all over again about what to do if something should go wrong during this period. If the cabin pressure is still normal, the capsule would still be in a "Go" condition for orbit—at least insofar as this particular criterion was concerned. If the cabin pressure should suddenly fall to zero fifteen seconds later, however, what do you do?

You have to decide fast. You ask all the flight controllers to report any small signs of trouble they may have spotted. Suit pressure, cabin temperature, fuel quantity, battery amps, body temperature, booster velocity, pulse rate, launch trajectory, respiration rate, watt-hours, CO_2, pressure-suit temperature, oxygen quantity. All are normal. All other systems are still "Go." You make the decision. You give the capsule a "Go" for one orbit. But only one. The reason for this is that the Astronaut must now rely heavily on his suit to keep him alive. Without pressure in the cabin, he has no other backup left—except one. Even if his suit should spring a leak and decompress, the pilot can still switch on his emergency supply of oxygen. This will be used up fairly fast under these circumstances, but it will be enough to last him for one orbit. According to our mission rules —and common sense—it would be the lesser of two evils to let the mission continue for one orbit rather than abort it at this point and subject the Astronaut to all of the rigors of high Gs, a shaky descent and a possibly hectic recovery that an abort at this critical point in the flight would entail.

This is the *kind* of decision that Chris Kraft has to be ready to make in a hurry on any morning that an Astronaut is going into orbit. Fortunately, no problems of this particular kind arose during the launching of MA-6. All systems remained "Go" throughout the launch. As Chris looked around the room, checking all of the display boards and glancing at each of us to see if we had any problems, I went on talking with John Glenn over the radio. This was my job, to make sure that the human system inside that spacecraft was still "Go." I could hear his voice clearly as he read off his meters and gauges to me. It vibrated a little here and there as the Atlas he was riding went through an early rough spot. We had expected that. But the timbre of his voice was good. He sounded exhilarated and in full command of the situation. I have a habit of raising my thumb when things are going well. It saves words. So when I knew Chris was looking at me as he glanced from one console and display board to another, I stuck out my left hand and raised it, thumb up. We were on our way.

The hours of preparation before the countdown were long and arduous for all the flights. Here, late at night under a wintry Florida sky, John Glenn and Scott Carpenter gaze into a plastic ball on which the position of the stars has been plotted. Carpenter, who was Glenn's backup pilot and helped him with his preparations, holds a flashlight as he orients the globe, and Glenn is about to look up from a constellation to see if he can spot it in the sky, just as he will on his flight in order to keep track of his position on the dark side of the earth.

One of the chores each Astronaut had to complete before he was ready was to draw up a detailed flight plan of his mission listing each event he was to perform or monitor during the flight and set it down in a precise time sequence that would match that of the actual flight. Here John Glenn spreads his own plan out on the bed of his motel room and sorts out the chronology until each event is listed in its proper place.

Safety was a paramount concern throughout the program. No effort was spared to erect special safeguards for every conceivable kind of emergency. Here, next to the towering Atlas that put John Glenn into orbit and jutting up to within a few inches of the capsule which rests on top of it, is a steel drawbridge that could be quickly lowered into position by remote control if anything occurred on the pad prior to launching that could endanger Glenn's life. Though the pad would be cleared of other personnel, Glenn could open his hatch, run across the drawbridge and ride down an elevator in the drawbridge tower to a waiting crash truck.

Weather was a constant problem before a launching. Here, in the early morning hours at Cape Canaveral, the searchlights bathe the Atlas gantry, and John Glenn is already locked into his capsule. But the clouds are thick and ominous, and later the same morning, after Glenn had spent more than five hours waiting for the go-ahead, the mission was canceled. He did not go into orbit until two months later.

On February 20, 1962, the weather was "Go," and so were all the systems. The steel gangplank had been folded back along the safety tower, and with a burst of flame the Atlas began to lift itself off the pad [OPPOSITE PAGE] with John Glenn's capsule perched on top. Glenn had problems during the four-and-a-half-hour flight. His controls did not respond as crisply as they should; and there was some anxiety as he re-entered the atmosphere about whether his protective heatshield was in its proper position.

But Glenn came through the flight in excellent shape. He sat inside the capsule as it bobbed about in the Atlantic awaiting recovery [ABOVE]. When he was lifted off the destroyer by a helicopter [RIGHT], it was evening, and he remarked that it was quite a thing for a man to be able to see four sunsets in one day.

Glenn returned to the U.S. for a tremendous hero's welcome. President Kennedy flew to Cape Canaveral to meet him. His family joined him there, too, and after a ceremony at which the President presented him with a medal, Glenn took his wife and children up to the capsule—which had also been brought back to the Cape—and showed them how well it had sustained the flight [BELOW]. Then there were more celebrations, including a thunderous welcome in Washington and the traditional ticker-tape parade up Broadway in New York [RIGHT].

THE

John H. Glenn, Jr.

MISSION

As the weeks went by from December and into February, it must have begun to look as if I were never going to go. We would have been ready at almost any time, but first bad weather, then troubles with the gear, then more bad weather kept piling up the delays. We were disappointed, of course, each time we had to stop. Scott and I were most concerned, however, about the 20,000 people who were directly involved with the program—the launching crews, the men in the recovery fleet, the technicians standing by at the tracking stations scattered around the world and all of the Mercury group who were on hand at Cape Canaveral. Each delay meant that these crews had to recycle their work, go through the long checklists and start up the countdown all over again. I think people normally build up to a peak when they are preparing for an event as complicated as this, and here we had a situation where we kept building up psychologically and nothing happened. It was like crying "Wolf!" over and over.

We need not have worried. The crews just kept working and they lost none of their zest or their sharpness. As for me, I was as disappointed as anyone, but I think the delays may even have been blessings in disguise. I do not believe a pilot who is preparing for a mission like this could ever reach the point where he could say to himself: "There is not

one thing I do not know about this job. I can do everything perfectly."
On the contrary, I think that if we had spent an entire year going from
one delay to another on this mission I would still have kept busy until the
last minute reviewing the flight plan, studying the charts and going over
each system to make sure that I could manage all of the complicated sequences. Readiness is relative, and we simply used the extra time to get
more ready. I went on studying, and we continued working on the trainers to practice our procedures. The men who were running the simulator
threw everything they could at us, including all of the failures they could
think of. You give conscientious technicians like this a challenge and they
will have at you. We repeatedly practiced maneuvering the capsule with
the manual controls. As it happened, this extra homework was very
useful.

A lot of people apparently began to worry about us when we had to
scrub one attempt because of weather after I had been in the capsule for
more than five hours. Some seemed to feel that I would have trouble
standing up through all of this waiting and suspense. One psychiatrist
who had never even seen me even ventured the opinion in public that it
would be impossible for me not to be suffering somewhat under the emotional strain. Well, if I was suffering I was not aware of it, and neither
were the psychiatrists who knew me well and whose job it was to keep
track of my emotions.

The nearest that I came to getting upset was after I visited a friend's
home in Cocoa Beach one evening for dinner. A couple of days later his
children had the mumps. I had never had them. Delays due to weather
and technical difficulties were facts that I could accept, but I had been
driving towards this goal for so long that the thought of going through a
postponement and getting a possible replacement while I recovered from
mumps would be hard to take. To make things worse, I woke up one
morning soon after this with a sore neck. I thought surely that the mumps
had arrived. Then I remembered that I had practiced making some strenuous head movements the day before with my helmet on. It was a drill for
an experiment which I would carry out in the capsule during weightless
flight, and I must have overworked some muscles in my neck. I also remembered that I had been exposed to mumps before. My wife had had
them when we were going steady in college; and my own children had
had them. Bill Douglas, our flight surgeon, felt that I might have had a
mild case at one time, which left me immune without my even knowing
it. At any rate, the threat passed. I was very careful from then on, how-

ever, to stay away from anyone who had the sniffles or even looked as though he might be ill. It was the season for colds and sore throats, and I had meetings to go to with people who had come from all over the country, so I tried to be careful and not expose myself any more than necessary.

Except for that one scare, the long period passed routinely. I continued with my physical training in order to stay in shape. I tried to run five miles each day until about a week before the launch date, when I cut this down to two miles. Three days before the launch I ran one mile. Then, on the last couple of days I rested physically and spent most of my time practicing on the procedures trainer. I also used the time to study the latest weather maps and review the star charts. There were medical examinations to take before each scheduled launch date, including a whole gamut of neurological tests and a complete physical that included chest X-rays, EKG readings of the heart, exercises to test my inner ear and my ability to keep my balance, and tests of blood and urine. These tests took most of a day. I spent half an hour one day breathing through tubes in my nose while technicians covered my face with plaster to make a cast. They used this to mold a plastic eyepatch which I planned to wear over my left eye during portions of the flight to help adapt my vision to the darkness.

A mountain of mail was piling up for me throughout this period, and I made a determined effort to answer it as it came in. At one point we had several file drawers and overflow boxes full of letters at Hangar S alone, with other mail pouring in at other NASA offices. Most of it was full of good wishes and prayers, for which I was very grateful. Some of the letters informed me that the postponements were God's way of letting us know that we should not be tampering with the heavens. I received one letter from a man who had seen my picture on the cover of *Life*. He said he could detect people's ailments by looking at their eyes and he thought I should know that I was suffering from gall-bladder trouble.

We were getting into February, and the mail was also heavy with valentines. One of them, I remember, read "I have SPACE in my heart for you." It came, I believe, from a young boy in grade school. One evening, after I had spent the better part of the day in the procedures trainer, I asked our secretary, Nancy Lowe, to stick around and help me whittle away at some of the mail. I had stocked the refrigerator in the aeromed area to provide for such late work periods, so we fed ourselves that evening by getting two cans of soup out of the hangar cupboard—chile con carne for Nancy and chicken noodle for me—and the cans of fruit cock-

tail and turkey from a couple of K rations. We had no dimes between us
to put into the milk machine, so we drank water and had some cookies
which Nancy had brought for dessert. We ate while we answered letters,
but it still made only a small dent. I am afraid we are still catching up
with this problem as this book goes to press. Perhaps I can pause here for
a second and use this opportunity to thank people once more for their
thoughtfulness. It was very much appreciated.

I rarely left the Cape during this period. About the only thing I could
not avoid was an occasional trip to Cocoa Beach for a haircut. On Sun-
days I went into Cocoa Beach to church. Occasionally Scott and I would
spend half an hour or so after dinner just listening to records. I cannot
really call myself an opera fan, but Bill Olsen, one of our assistant dieti-
cians, had a stereo record player, and I must have worn out one favorite
of mine, Puccini's *Madame Butterfly*. During one long postponement, I
had a good weekend in Arlington with my family. They understood as
well as anybody the reason for the delays, but it did us all good to see
each other again. While I was there, President Kennedy asked me to stop
by briefly at the White House. He expressed his best wishes, and I was
very grateful for his support and understanding.

On the morning before the launch, I studied the results of the first part
of the Atlas countdown, which was already in progress. The technicians
had already run power through the capsule and they were checking the
oxygen equipment and the telemetry signals. They were verifying that
all of the cameras and radar beacons and command circuits were working,
and they were hooking up the igniter for the rockets on the escape tower.
I went into Cocoa Beach for one final haircut, but I was not too optimistic about our chances. I still felt the same way that evening. All of the
booster and capsule systems were "Go," the tracking stations and recovery ships were "Go," the downrange weather was good and I assured
Walt Williams at the scheduling meeting that I was "Go." But there was
a local storm front moving across Canaveral at the time, and that night
the weather people gave us only a fifty-fifty chance that it would clear in
time for a launch the following morning. These odds had never taken us
very far on earlier tries, so I was not exactly hopeful. It seemed more
likely that we might have to wait at least one more day. I called Annie
and the children and then my parents in New Concord and told them not
to be disappointed if we had another delay. Then I wrote Nancy Lowe a
note and left it on her desk with my car keys and a check which would

cover small expenses I had run up for meals and laundry and things like that. I got to bed about seven.

I woke up about 1:30 in the morning, half an hour before I was scheduled to get up. I had slept fine, and I just lay there for a while running mission procedures over in my mind. It still did not feel too likely for me that morning, and when Bill Douglas, our doctor, came in to see if I was awake he verified that the weather was still fifty-fifty. We have double-decker bunks in the crew quarters at Hangar S, and for some reason I always like to sleep on the top bunk, even if the bottom one is vacant. Bill walked over and rested his arms on the edge of my bunk while we talked quietly. The count was moving right along on schedule, he told me, but there was still a solid cloud cover in the launching area. Scott Carpenter got up at midnight and went out to the pad to monitor the capsule checkout and run a final test on the controls. He called back to say that everything was ready. The booster was in good shape. All we needed now was a break in the local weather. Like the fine doctor that he is, Bill just eased into the question: How did *I* feel? I assured him that I was ready and jumped down from the bunk to shave and shower.

I put on a terry-cloth robe and sat down with Bill Douglas, Deke Slayton, Walt Williams and Merritt Preston, chief of the preflight operations division, for a good breakfast of orange juice, eggs over easy, filet mignon, toast and Postum. Deke was scheduled to make the next orbital flight, so he was looking over the course that morning for future reference. After breakfast Bill and several other specialists gave me a final examination and started to attach the biomedical sensors on my body. We checked them to make sure they would work properly and would transmit information back to earth on my heart status, blood pressure and temperature.

The official report of this examination stated that John Glenn was found to be a "calm, healthy and alert adult male. Vital signs were as follows: pulse, 68 beats per minute and regular; blood pressure, 118 over 80; respiration, 14 breaths per minute; oral temperature, 98.2 degrees Fahrenheit; and nude weight, 171 pounds 7 ounces. Eyes, ears, nose and throat were normal and unchanged from previous examination. Lungs were clear throughout."

When the sensors were in place, I started to climb into the pressure suit. As usual, Joe Schmitt, our suit technician, gave me a hand. Joe is one of the most conscientious and hard-working men I know. He has presided over all of our suitings-up, and he is the kind of dedicated citizen and skillful craftsman who makes Project Mercury go. To make sure that the air we used to test out the pressure suit was pure, Bill Douglas had rigged up an experiment of his own. He ran a small hose from the main air supply tube and stuck it into a bowlful of tropical fish on his desk and let it bubble away night and day. The idea was that if any bad air got into the system, Bill's fish would react to it and we would know we shouldn't use the air. While Joe was checking the suit for leaks, I casually asked Bill if he realized that two of his fish were floating belly-up in the bowl. Bill rushed over for a close look before he discovered that his fish were fine and that his patient was only joking.

It was time now to get into the van and ride out to the pad. All of my maps and instruments were already on their way, so all I had to carry was the portable air-conditioner which would keep me cool inside the suit until I could plug into the capsule system. There were about 150 people clustered outside the hangar when I came out and walked through a roped-off area towards the van. This was a pretty big crowd for five in the morning. There was some quiet hand-clapping as I walked past, and I waved back my thanks. We climbed the steps into the van, and I settled in a reclining contour couch for the short trip to Pad 14. Deke and Bill Douglas and a few others rode along with me. Ken Nagler, one of the weather experts, was already in the van marking the latest information on the maps, and he briefed me most of the way. Joe Schmitt continued to check on my suit, and we plugged the biomed sensors into a device on the van to make sure that they were still functioning. As we got close to the guarded gate that leads into the launching area, there was the inevitable joke about whether we all had our security passes with us to get on the pad. There was no place to pin a badge on my pressure suit, of course, so someone joked that I probably would not be allowed in. The security guards waved the van right through the gate, just as they always do on these occasions, and as we pulled up to the pad we opened the blinds on the van windows, and I could see the gantry. It was a beautiful sight. It was still dark, but the big arc lights shone white on the Atlas booster. It looked like something out of another world.

We stayed in the van during two holds. One of them was planned, in order to make a final check on the weather before I started climbing into the capsule. The blockhouse called for the second hold to replace an item in the guidance system of the Atlas which was not checking out properly. This took 45 minutes. I went on reviewing weather data and studying the flight plan. Finally, at one minute before six, we all walked to the elevator which was waiting for us at the foot of the gantry. The crews clapped and waved their good wishes. I nodded my thanks, and felt very good as I realized how much they were with me. Everyone is a teammate with you on a morning like this. Then I stepped into the elevator and rode up to the eleventh deck—spacecraft level.

Everyone up there seemed ready for me to get in without further delays. Scott Carpenter had completed his checks on the capsule and came over to tell me that everything looked good. The weather people were now forecasting a possible break in the cloud cover by midmorning. This would allow the visual tracking and photography of the launch which the technicians required, and which was all we needed now. There was a different atmosphere on the gantry. There was less casual talking among the crews than you would hear during a dress rehearsal. Everyone seemed to sense that we were going for real this time, and in keeping with the seriousness of the occasion they became more purposeful and business-like than ever.

Scott was standing off to one side, out of the way, and I walked over to him for a moment. I know something about being a backup pilot, having done the job twice myself. It is hard work, and the personal satisfaction is limited to helping someone else. Scott had pitched in from the beginning as if he were preparing for his own flight, not someone else's, and I appreciated this very much. Scott was my alter ego. He took an enormous load of detail off my mind and left me free to concentrate on areas where I thought my attention was most needed. He attended many meetings that both of us could not possibly have attended, and the decisions he made were the same as mine would have been, without exception. All during this period he had to keep himself in perfect shape and trained to a fine edge so that he would be ready to make the flight if I were unable to go at the last minute. Scott is much more to me, however, than a willing colleague. He is also a close friend, and that morning we both felt emotions which could not be expressed when we shook hands and he wished me good luck.

It was now T minus 120 minutes on the count and time for me to get in. It was a tight squeeze. It always is when you have your bulky helmet and suit on. I stuck my left foot over the sill first, and then inched the rest of my body under the instrument panel and into the couch. Joe

Schmitt leaned in after me to put the restraining straps on loosely. We would fasten them later. Then we plugged the suit into the environmental control system. Just as we were getting ready to run a pressure check on the suit, I pulled the left microphone down inside my helmet to adjust it. I had done this a thousand times, but this time the microphone support broke and dangled down inside my neck ring. I was wearing two mikes inside the helmet, one of which was only a backup for the other. But since we had planned the whole program around the principle of having alternate systems on hand in case of trouble during a flight, we did not want to launch with only one lip-mike. We sent down to the van for another support from my backup helmet, which we kept there, and Joe Schmitt crawled in on top of me to put it in place. It was difficult work, for he had very little room to work in. Joe looked just like a dentist as he fiddled around in front of my face with his fingers and kept moving a small, undentist-like wrench back and forth in front of my lips. Joe is very good with his hands, however, and he had the trouble repaired in about 10 minutes.

I closed the faceplate on my helmet; the oxygen started flowing; we purged my suit of air and checked the pressure in it. Then we conducted a cockpit check to make sure that every switch was set properly and that nothing had been left ON that should be OFF—or vice versa. I had to do the checking on the instrument panel. I checked as the technicians called out the switches. These checks were lengthy, perhaps lengthier than they had to be in order to give the pilot plenty to do. This was not necessary, however. To be sure, I was aware of some apprehension. It is normal, I think, for humans to have a certain amount of fear of any unknown situation. But the best antidote for this is to know all you can about the situation. I had done this by training. During the years of preparation and the final months of constant practice and study, the unknown areas had shrunk to what I felt was an acceptable level. So far as I was concerned, if I was so shook up that I had to keep busy to stay calm, I did not belong here.

About an hour after I climbed in, the crew started to put the hatch on. This is an interesting moment, as Al Shepard and Gus Grissom had discovered for themselves, when things really begin to come home to you. Up until now, people have been reaching into the capsule with their hands to fix something behind you, pat you on the shoulder or shake your hand. You know how strongly they all feel, and you feel strongly, too. But there is no real feeling yet of being on your own. Then, suddenly, there are no more hands. The hatch is closed. It is quite a moment—and a good one.

We had to go through this moment twice, however, for as the technicians locked the hatch in place, one of the seventy bolts which hold it onto the capsule broke. We had to replace the bolt and start all over again. This took about 40 minutes, and I began to be a little concerned. The weather was beginning to show definite signs of clearing by this time, and I was afraid that we might wind up with a clear sky and have to scrub the mission because of the hatch.

The bolt was finally replaced, however, and the hatch was secured. This was just before eight o'clock, and a few minutes later the loud horns began to blow, warning everyone off the gantry. Through the periscope, I could see people leaving. One of the last to depart was Bill Douglas. He came over to look in through the scope and gave me a smiling salute.

We were in the last hour of the count now, and I was talking continuously with Scott, who was back at the blockhouse. At one point during the count, when most of the activity centered around the booster operations and I had no scheduled duties in the capsule, Scott was thoughtful enough to put through a call to Annie for me. While it was going through on a private channel, the gantry opened up enough so that I could look out towards the sea through the periscope and see a patch of blue through the clouds. When Annie came on the line, I told her about the sky, and described some of the things that were going on inside the capsule while she saw some of the preparations that were going on outside as she watched them on TV a thousand miles away. I had wanted to buck her up, but she sounded firm as a rock as she told me how the children were and what was happening in our home in Arlington. I also talked to Dave and Lyn. It was not only a welcome break in the preparations, but a time of closeness for all of us that I will always remember. I felt confident that Annie was in good shape, and I wondered as she hung up who had bucked up whom.

The Atlas is an interesting thing to sit on top of when the gantry is gone. You are 74 feet above the ground, and the booster is so tall that it sways slightly in a heavy gust of wind. In fact, I could set the whole structure to rocking a bit myself, just by moving back and forth in the couch. I could look at the whole sky now as it was filling with bigger patches of blue. Through a mirror which was mounted near the window —like the rear-view mirror on a car—I could see the blockhouse where the technicians were monitoring the countdown and on across the Cape to other installations where we had prepared for the flight. Through the periscope I looked out at the Atlantic and eastward along the track that I would soon follow. I took only a few brief glances outside, however, for I was still checking switches and monitoring the systems on the instrument panel in front of me. If the oxygen-pressure or fuel-supply gauges should suddenly drop, indicating a failure somewhere in the system, I wanted to know about it instantly. Everything was proceeding normally, however. I could also hear some of the telltale sounds over the noises in my headset. I could hear the pipes whining and crackling below me as the liquid oxygen flowed into the booster tanks, and I could hear a vibrant hissing noise as the tanks were supercooled by the cold lox. The metal walls of the tanks set up a high-frequency resonance and buzzing as they shivered from contact with the oxygen. When the technicians gimbaled the engines for a moment to make sure they were working, I could hear a faint shaking and thumping coming up through the booster. These were all sounds that I had been briefed about, and it was reassuring that each new sound indicated that we had reached another stage in the countdown. I talked over the hard line with Paul Donnelly, the NASA test conductor who was in charge of preparing the capsule system for launch, and with Tom O'Malley of General Dynamics, the firm that made the Atlas, who was in the blockhouse as the over-all test conductor. I also made radio checks with Al Shepard, who would be communicating with me from the Mercury Control Center during the flight and was keeping me posted on the latest status of the weather, the tracking range and the emergency recovery vehicles stationed on the Cape. There was a real feeling of excitement in many of the voices—"Go" fever, some people have termed it—and this was heightened at T minus 35 minutes when the decision was made to start putting in the final load of lox. This was one of the landmarks on the countdown, and when I heard that this procedure was beginning I knew that we were almost on our way.

I had to wait through two more brief holds, however. One came at T minus 22 minutes when one of the loxing valves stuck and the technicians had to shift to a smaller valve to complete the loading process. This was a minor problem and it took about 25 minutes. Then, at T minus 6 minutes and 30 seconds, there was a two-minute hold to check on a power failure in the computer system at the Bermuda tracking station. Bermuda was a key station in the network because it lay close to the area where I would actually be going into orbit and served as a backup control center for Cape Canaveral. I knew that I would not be launched unless this station was in perfect shape. I was greatly relieved, therefore, when the power was restored and the count was resumed.

We were getting down to the short rows now. Over the radio I could hear the people responsible for each of the systems reporting in to the test conductor. "Communications, 'Go,' " "ASCS, 'Go,' " "Aeromed, 'Go,' " "Range, 'Go.' " The Astronaut was one of the last items on the list, and when my turn came I said, "Ready." I was.

About a minute and a half before lift-off, I did a few quick exercises to make sure that my body was toned and ready for the launch. The aeromed people asked for one final blood-pressure check before lift-off. They had been asking for this all through the count, and pushed a button which started the recording instrument and I pumped up the bulb, took the blood-pressure reading automatically from a cuff on my left arm and sent it along by telemetry to the Control Center. Then, I put my left hand on the abort handle, as procedure requires. At T minus 35 seconds, a special countdown started for dropping the umbilical cord which had been providing external power and cooling for the capsule up until now. This was the last physical link between the capsule and the ground, and I watched through the periscope as the umbilical fell away and I heard it fall with a loud plop. The periscope retracted automatically, and this shut off my view from that direction. The land lines to the blockhouse and Control Center were cut off now and we communicated from now on only by radio. I could detect a tone of excitement in the voices in my headset, and as the countdown we had practiced so often ran down for this final time, I shared the feeling. At T minus 18 seconds there was a planned momentary hold of two or three seconds while the automatic engine starter was switched on. I did not hear it at the time, but just as the engine sequence started and he knew this was it, Scott spoke into his microphone in the blockhouse. "Godspeed, John Glenn," he said. I heard it later on a recording of the transmissions; it was a very impressive moment. No one would push any more buttons or take any further positive action now—except to stop the show at the last second in case of an emergency. Then Al Shepard's voice gave me the final 10 seconds of the count. He reached zero and the engines started.

I could feel the engines light off as the capsule vibrated from their ignition, and I could hear a faint roar inside the capsule. The booster stood fast on the pad for two or three seconds while the engines built up to their proper thrust. Then the big hold-down clamps dropped away and I could feel us start to go. I had always thought from watching Atlas launches that it would seem slow and a little sluggish, like an elevator rising. I was wrong; it was not like that at all. It was a solid and exhilarating surge of up and away. Al Shepard received a signal that I was lifting off and confirmed it for me over the radio. The capsule clock started right on time and I reported this.

"The clock is operating," I said. "We're under way."

It was 9:47:39 A.M. (E.S.T.) when the Atlas left the ground. Al rogered for the message and told me to stand by for the 20-second count to start up the stop watch I had on my right wrist as a backup timer. It was already preset at 20 seconds; all I had to do was take my left hand off the abort handle for a second or two to push in the stem. The personnel down in the Control Center had a lot of fast figuring to do now as the tracking instruments and telemetry circuits started sending in data which had to be analyzed in a hurry to determine exactly how well the launch was progressing.

The launch itself was the first of four hurdles that I had to jump to get properly into space, and it was a big one. The booster had to function perfectly, and it had some maneuvers to start executing immediately. Pad 14, where we took off, is lined up with the Atlantic Missile Range which lies to the southeast of the Cape—at an azimuth, in fact, of 105 degrees. In order to get into the correct orbit, however, the booster and the capsule had to start off almost immediately on an azimuth of approximately 75 degrees. For the first two seconds, the Atlas went straight up. Then, for the next 13 seconds, the automatic guidance system which was built into the booster made it roll gradually to a northeast heading. Eight seconds after lift-off I reported that this was taking place.

"We're programing in roll O.K.," I said. I could feel the motion and see it happening by glancing out the window through the mirror.

Five seconds later, I reported that the flight was getting "bumpy along about here." This was something that we had predicted by studying the pattern of previous Atlas launches, and it was nothing to worry about. It was just that I could feel a little resonance or roughness and wanted the Control Center to know for the record what was happening. I found out later that my voice was also vibrating slightly over the radio as I called in.

We went over that first hurdle in good shape. I started the backup clock on my wrist at T plus 20 seconds. Then I started ticking off a list of items—oxygen and fuel supply and the ampere reading on the batteries. Everything read off just as it should; the capsule was functioning perfectly. Al Shepard came on the radio to assure me that from the telemetry indication in the Control Center the flight path looked good.

The second hurdle came when I started into what we call the "maximum q" area, or the portion of the flight where we knew we would encounter the highest aerodynamic forces against the capsule and the booster. This comes at an altitude of about 35,000 feet and is an area where we had had some difficulties earlier in the Mercury program. We entered the area about 45 seconds after launch. I reported to Al that I could feel the vibration building. This phase lasted about 30 seconds. The shaking was more pronounced at this point. I did not expect any trouble, but we knew there were certain limits beyond which the Atlas and capsule should not be allowed to go. One Mercury flight with an empty capsule had ended here when the Atlas blew up. The abort system had worked fine, however, and the capsule came down in such good condition that the engineers were able to use it later on another test flight. Structural changes had been made in the Atlas and tested on later flights, and, in any event, automatic sensors in the ASIS system would abort the mission and break me loose from the Atlas if the vibrations got too high. Nevertheless, it is difficult for the human body to judge the exact frequency and amplitude of vibrations like this, and I was not sure whether we were approaching the top limits that ASIS was set for or not. As it turned out, the capsule was under aerodynamic pressure of 982 pounds per square foot during this phase of flight. This was well within limits, and we made this hurdle, too. I saw what looked like a contrail float by the window. I reported again on the supplies of consumables that we had stocked in the capsule—fuel, oxygen and battery power. At 1 minute and 16 seconds after launch, Al confirmed that I had passed through "maximum q," and I answered that I felt good and that the flight was "smoothing out real fine." The G forces were building up to about 6 now. I strained against them to make sure that I was in good shape. I was.

At 2 minutes and 11 seconds after launch we jumped the third hurdle right on schedule when the two big outboard booster engines shut down and dropped away. We were out of the atmosphere by now, and had built up enough speed so that all we needed was the long, final push from the sustainer engine to drive us into orbit. There was no sensation of speed, however, because my window was not aligned so I could see anything and use it as a reference point. The only time I took my left hand off the abort handle during this period was to be ready to jettison the escape tower at 2 minutes and 34 seconds after lift-off in case it failed to leave automatically. Now that we were out of the atmosphere where the drag forces were high, we no longer needed the escape rocket to force us free of the booster; and since the tower's extra weight would just waste fuel that we needed for getting into orbit, it was supposed to jettison and drop away. I thought I saw it go about 20 seconds early, and I reported this to Al. I did not actually see the tower, but I did see smoke go by the window and I assumed it was from the escape rocket firing. I was wrong, however. The smoke had apparently been deflected around the capsule during booster engine shutdown and staging. The tower went on schedule a little later and shot out a momentary cloud of flame and smoke. I could feel a slight bump as it took off, and I watched it going straight away from me, accelerating at a tremendous clip. It disappeared quickly. The G forces on me had dropped at staging to about 1.5.

The sustainer engine was being guided now from the ground, and the Atlas was completing another precise maneuver. It had been turning a corner in the sky and programing over on its pitch axis at the rate of about 2 degrees per second until at one point—just before the tower fired —the capsule was actually riding lower than the engine section of the booster. The booster-capsule combination was still climbing, but it was changing course through space at this point so that it would thread itself through the keyhole and not keep going straight up. I caught a quick glimpse of the ocean through the window as we pitched down. I knew we would pitch up again and make the approach to the fourth hurdle, the insertion into orbit.

The insertion was perfect. Al and I had been talking during the corner-turning stage. He said that the Cape was "Go" and was standing by for me. "Roger," I said. "Cape is 'Go' and I am 'Go.' Capsule is in good shape." Then I read off the instrument readings again and said that all systems were "Go." Al said we had "twenty seconds to SECO"—or sustainer engine cutoff. At 5 minutes, 1.4 seconds after lift-off, the sustainer engine shut down. Then the bolts which held the booster and the capsule together exploded, and the posigrade rockets fired to push the capsule away from the booster. I could hear and feel each of these explosions take place. There seemed to be a barely noticeable sensation of tumbling forward when the capsule separated; it was only momentary and I did not feel disoriented at any time as we turned around.

This was it. We were 100 miles up and going at a velocity of 25,730 feet per second. I went weightless as the G forces dropped from 6 to zero; it was a very pleasant sensation. The periscope extended, and the capsule began to turn around automatically to orbital attitude—blunt end forward—which it would hold throughout the three orbits. The automatic system accomplished this maneuver by activating the nozzles which then turned the capsule. Now, for the first time, I could look out the window and see back along the flight path. I could not help exclaiming over the radio about what I saw. "Oh," I said, "that view is tremendous!" It really was. I could see for hundreds of miles in every direction—the sun on white clouds, patches of blue water beneath and great chunks of Florida and the southeastern U.S. Much nearer, I could see the Atlas drifting along by itself, about 200 yards behind me and slightly above. But these were only quick glances, for I had to concentrate on monitoring the capsule and preparing to start on the flight plan items. I was still not sure about our insertion conditions. Then Al Shepard called with the message that I had been waiting for. "You have a 'Go,'" he said "for at least seven orbits." I was really jubilant. Al meant that the computers at the Cape had run through all of the data and had indicated that the insertion of the capsule was good enough for a minimum of seven orbits. This is more or less a standard computer figure. The factors would probably have been good enough for seventeen or seventy orbits if we had been able to carry enough fuel and oxygen for such a long mission.

I loosened my chest strap now and went to work. Until this moment, every sequence and event had been extremely time-critical. That is, the successful jumping of each hurdle had depended on split-second timing. Now that I was in orbit and at zero G, however, we did not have to be quite so conscious of each fleeting second until it was time for firing the retro-rockets at the start of re-entry. I had a lot of work to do, and in order to get it all done I would have to adhere to the schedule as closely as possible. But at least I did not have to sit on pins and needles every second. Nancy Lowe had typed the flight plan for me on a long piece of paper which we rolled up into a tiny scroll which I would unroll a bit at a time. I did not discover until later that Nancy had typed a private message at the end of it to give me something pleasant to think about when I saw it at the end of the mission. "Don't get too much sun on GTI," she wrote. This referred to Grand Turk Island, where I would proceed for the debriefing and medical examinations after the mission— assuming it was a success. It was nice to know that Nancy was so sure of the outcome.

The plan called for me to spend most of the first orbit getting used to the new environment and helping the ground stations establish an accurate pattern of radar tracking so they could pin down an exact orbital path early in the mission. I was also scheduled to test out the various systems on the capsule before we got too far along, so that we would know if we had any problems before we committed ourselves to a second orbit. Eight minutes after launch, as soon as Gus Grissom finished relaying to me the correct times for retro-firing from his Cap Com desk at the Bermuda tracking station, I started checking out the attitude control systems —automatic, manual and fly-by-wire. I tried each system on all three axes —yaw, pitch and roll—and in all directions—up, down, left and right. This took about two minutes, and by the time I was finished with it I was almost across the Atlantic and was in communication with the tracking station in the Canary Islands. All of the controls had responded perfectly, just like clockwork. The stick handled very well. I was happy to see this, for there is always some doubt whether such complicated controls will work as well under actual conditions as they do on the procedure trainers —and whether I would be able to work as well, also. I could see no difference—at least not yet.

Inside the spacecraft, I could hear a number of muffled sounds. There was some noise from the gyros which gave us our attitude references, another noise caused by the inverters which were converting D.C. power into A.C., the hiss of the oxygen flow as it ran through the hose in my helmet, and sounds from the nozzles which were spitting out hydrogen peroxide to correct the attitude of the capsule.

In addition to closely monitoring all the systems, I started making a few observations out the window. Since I was facing backwards, everything came out from underneath me, similar to the way things look when you ride backwards in a car, and it seemed to move more rapidly than I had thought it would. The sense of speed was similar to what you normally experience in a jet airliner at about 30,000 feet when you are looking down on a cloud bank at low altitudes. I think our training devices had been a little inadequate on this score; they had given us less sensation of motion and speed than I could feel in actual flight. Just before I finished crossing the Atlantic, I had my last glimpses of the Atlas. It was still in orbit, about two miles behind me and a mile beneath me. It was bright enough so that I could see it even against the bright background of the earth.

I saw the Canary Islands through the periscope and then saw them through the window. They were partially hidden by clouds. While I was reporting in by radio to the Canary Island tracking station I had my first glimpse of the coast of Africa. The Atlas Mountains were clearly visible through the window. Inland, I could see huge dust storms blowing across the desert, as well as clouds of smoke from brush fires raging along the edge of the desert. One of the things that surprised me most about the flight was the percentage of the earth which was covered by clouds. They were nearly solid over Central Africa and extended out over most of the Indian Ocean and clear across the Pacific. I could not establish the exact altitude of all of the various layers, but I could easily determine where one layer ended and another layer began by the shadows, and I believe that with better optical instruments we can contribute a good deal to the art of weather forecasting from this orbital altitude.

I gave a lengthy report to the Canary Island station on the status of the various capsule systems. I was still maintaining a careful orbital attitude here so that the tracking radar could get a good reading on our altitude, range and position. All of the capsule gauges had normal readings, and the control system was still operating very well. The Canary station asked me for a blood-pressure check. I pushed the button, pumped up the cuff, and a minute later the station reported that it had received the telemetry message and that my blood pressure reading was 120 over 80. The station was also able to tell me, from reading telemetry reports which were emanating from the capsule, that the temperature on my automatic fuel line was 70 degrees and that the temperature of the manual line was 100 degrees. Both of these readings were within limits and I rogered for the message.

Four seconds after I lost radio contact with the Canary Island station, I could hear the Cap Com at the Kano station in Nigeria calling me. I read off the figures on fuel, oxygen and cabin pressure, and then I opened up the faceplate on the helmet for a few seconds to take a xylose pill. This is a special sugar tablet which allows the doctors to determine some things about how well the digestive system is functioning.

I carried out a yaw maneuver while I was in contact with the Kano station, so I could report more fully on the status of the control systems. I used the fly-by-wire mode, which, as Al Shepard has explained earlier in the book, utilizes a combination of the manual control stick and the hydrogen peroxide nozzles which would normally be activated by the ASCS or automatic system. This method allows you to override the automatic system manually and still make use of the supply of fuel which it has on hand. The system worked fine. Before I left the area, the Kano Cap Com informed me that most of this part of Africa seemed covered with dust at the time, which was precisely how it looked to me.

Zanzibar was the next tracking station, and here the flight surgeon who was on duty came on the air to discuss how I was doing physically. I gave him a blood-pressure reading, but before I pushed the button and pumped up the bulb I pulled thirty times at the bungee cord which permitted me to exercise with a known workload that could be compared with the same exercise taken on the ground. The cord was attached under the instrument panel, and I gave it one full pull per second for 30 seconds to see what effect exercise would have on my system under a condition of weightlessness. The only real effect it had on me at the time was the same effect it had had on the ground. It made me tired. My pulse went up from 80 beats per minute to 124 beats in 30 seconds, but it returned to 84 beats per minute within a couple of minutes. My blood pressure read 120 over 76 before the exercise period and 129 over 74 afterwards. This was the sort of mild reaction we had expected from doing similar tests on the procedures trainer. The doctor also asked me what physical reactions, if any, I had experienced so far from weightlessness. I was able to tell him that there had been none at all; I assured him that I felt fine. I had had no trouble reaching accurately for the controls and switches. There had been no tendency to get awkward and overreach them, as some people had thought there might be. I could hit directly any spot that I wanted to hit. I had an eye chart on board, a small version of the kind you find in doctors' offices, and I had no trouble reading the same line of type each time. After making a few slow movements with my head to see if this brought on a feeling of disorientation, I even tried to induce a little dizziness by nodding my head up and down and moving it from side to side. I experienced no disturbances, however. I felt no sense of vertigo, astigmatism or nausea whatever.

In fact, I found weightlessness to be extremely pleasant. I must say it is convenient for a space pilot. I was busy at one moment, for example, taking pictures, and suddenly I had to free my hands to attend to something else. Without even thinking about it, I simply left the camera in mid-air, and it stayed there as if I had laid it on a table until I was ready to pick it up again. The fact that this strange phenomenon seemed so natural at the time indicates how rapidly man can adapt to a new environment. I am sure that I could have gone for a much longer period in a weightless condition without being bothered by it at all. Being suspended in a state of zero G is much more comfortable than lying down under the pressure of 1 G on the ground, for you are not subject to any pressure points. You feel absolutely free. The state is so pleasant, as a matter of fact, that we joked that a person could probably become addicted to it without any trouble. I know that I could. The only catch that I can think of on a space flight is that you would have to be careful about the kind of food you carried along. Cookies or crackers or anything crumbly could be a nuisance, for the pieces would float around and get in your way. But I think that an Astronaut could easily take a plain old ham sandwich up with him, complete with mustard, and not have to rely on tubes of vegetable purée and apple sauce—although most of them taste good, too.

We did have to go back to the drawing board after my flight, however, to devise better ways of carrying miscellaneous items and loose instruments in a weightless state. The items that I took along included a camera, filters, rolls of film, a photometer, a pair of binoculars and a special instrument which had been designed for us by Dr. Robert Voas, our training officer, and was capable of making a number of astronomic and physical measurements outside the capsule. It was christened as an EX-TINCTIONSPECTROPHOTOPOLERISCOPEOCCULOGRAVO-GYROKYNETOMETER. The syllables were in honor of the many different functions which can be separated out of that long name, but we called it the "Voas meter" for short. We knew ahead of time that any one of these pieces of equipment could get out of hand if we ever let go of it, so we designed a case to serve as our ditty bag and anchored all of the equipment inside the bag, each piece on its own three-foot length of string. The bag was next to my right arm. We assumed that we could take out one piece of equipment at a time, use it, then put it back again when we were finished with it. The cord acted as a retrieving line in case the object floated away. This proved to be not so easy in practice, however, and by the time I had used the items through the mission the lines were very tangled. Putting everything back in the ditty bag was like being a space-age Pandora. The principle was sound, however. I reached in once for a roll of film which was *not* tied down by string. I let it float for a minute while I opened the camera. Then I accidentally hit the roll and it went floating out of sight behind the instrument panel. I never did retrieve it.

By the time I had completed various medical tests for the flight surgeon at Zanzibar, I had also completed preparations for the coming of darkness. We had hoped to make some observations of the moon and stars from orbital altitude during the three 40-minute nights which I would experience on the flight, in order to determine how visible the horizon would be at night and how useful it might be for maintaining the capsule's attitude. This meant that I had to be prepared for night vision—or "dark adapted"—well before night came. I placed red covers over the lights in the cockpit and turned off the special photo-lights which helped us make pictures of my reactions during the flight. I turned on the tiny bulbs which we had had installed at the ends of the index and middle fingers on each glove. These served as miniature flashlights and were very useful. I also tried to install the special eyepatch which had been molded to fit the left side of my face. The patch did not work very well, however. My face was moist and the special tape we had brought along failed to keep the patch in place.

I witnessed my first sunset over the Indian Ocean, and it was a beautiful display of vivid colors. The sun is perfectly round and it gives off an intense, clear light which is more bluish-white than yellow, and which reminded me in color and intensity of the huge arc lights we used at the Cape. It was so bright that I had to use filters to look directly at it. Then, just as the sun starts to sink into the bright horizon, it seems to flatten out a little. As the sun gets lower and lower, a black shadow moves across the earth until the entire surface that you can see is dark except for the bright band of light along the horizon. At the beginning, this band is almost white in color. But as the sun sinks deeper the bottom layer of light turns to bright orange. The next layers are red, then purple, then light blue, then darker blue and finally the blackness of space. They are all brilliant colors, more brilliant than in a rainbow, and the band extends out about 60 degrees on either side of the sun. It is a fabulous display. I watched the first sunset through an instrument we call a photometer, which has a polarizing filter mounted on the front of it so you can look directly at the sun without hurting your eyes. I discovered later that it was possible to look directly at the sun without the photometer, just by squinting my eyes, the same as we have always done from here on the surface of the earth. We had thought that perhaps it would be too bright for that above the atmosphere.

I saw a total of four sunsets before the day was over—three during the flight and a final one after I had landed and been picked up by the destroyer. Each time I saw it set, the sun was slightly to my left, and I turned the spacecraft around a little on its yaw axis to get a better view. One thing that interested me was the length of the twilight. The brilliant band of light along the horizon was visible for up to five minutes after the sun went down, which is a long time considering the fact that I was moving away from the sunset and watching it occur at eighteen times the speed at which we normally watch sunsets from down here on earth.

Then the earth was dark; looking down at it was like gazing into a black pit. It was bright again, however, as soon as the moon came up. The moon was almost full. The clouds below showed up clearly in the moonlight, and I was able to estimate my angle of drift by looking down at the formations far below me.

I was able to see the horizon at night, and this enabled me to correct the attitude of my spacecraft against the horizontal plane of the earth below. I noticed an unexpected effect along the horizon on the night-side. There seemed to be a layer of haze about 2 degrees thick which hung about 6 to 8 degrees above the real horizon and lay parallel to it. I first noticed this phenomenon over the Indian Ocean as I was watching the stars set. They became dim for a few seconds as they approached the horizon and then they brightened again before they finally went out of sight. I looked carefully, and there seemed to be a definite band of some kind where they dimmed. It was not white like the moonlit clouds, but more tan or buff in color. And it did not have a definite configuration. The only real sign that a layer of some kind was there was that the stars dimmed as they passed through this area and then brightened again. The same phenomenon occurred on all three orbits, and it was most notice-able when the moon was up.

I had thought that I might be able to see more stars than I did. I was prepared to study the various constellations and groupings and count the number of stars in each one to see if I could spot any that we do not normally see clearly from beneath the atmosphere. I did see a whole sky full of stars, and it was a beautiful sight. The effect was much the same as you would have if you went out into a desert on a clear night and looked up. It was not much more than that, however. Laboratory tests of the window made before the flight were correct. The heavy glass in the window provided about the same attenuation that the atmosphere does, and though I saw the stars clearly—and they did not twinkle—I saw about the same number as you would on a clear night from earth. We will probably have to have a new kind of window before we can do much better. I did see a few stars during the day, shining against the black sky. But they were far more clear at night. The stars that I saw at night were of some help in delineating the horizon so I could control the attitude of the capsule. The constellations of Orion and the Pleiades were especially bright, and on one pass over the Indian Ocean I focused on Orion through the center of the window and used it as my only reference for maintaining attitude. All in all, a night in space is a beautiful sight. You see the moon shining bright on the clouds far below you and fields of stars silhouetting the horizon for hundreds of miles in each direction. Just like the sun, the moon and the stars declined and finally set at a speed eighteen times faster from my fast-moving, orbital vantage point than they do for us here on earth.

I saw my first and only signs of man-made light as I came over Australia on my first pass. Looking through the window, I could see several great patches of brightness down below. Gordon Cooper, who was on duty as the Cap Com at the tracking station in Muchea, Australia, had alerted me to look off to the right. He knew that the citizens of Perth and several other cities and towns along the coast had turned on all of the lights they had as a greeting, and when I spotted them I asked him to thank everyone for being so thoughtful. I gave Gordon a reading on fuel, oxygen and amps and told him that the control system was operating in fine shape. I also told him about the haze layer that I had seen, and said that I had the Pleiades in sight as we were talking. Gordon said I should be picking up Orion and Canopus and Sirius very shortly. He also relayed to me the information that the launching had been so near perfect that the orbital velocity was only 8 feet per second under what we had predicted it would be. In other words, it was 25,730 feet per second instead of 25,738 feet per second. When you are dealing with figures like that, worrying about a difference of eight is like quibbling over one drop of spilled milk out of a barrel.

"That was sure a short day," I said to Gordon.

He didn't hear the remark and asked me to repeat it.

"That was about the shortest day I've ever run into," I said.

He heard me that time, and then we went back to work. Cooper said the surgeons were standing by for a new blood-pressure reading, so I pushed the button and pumped up the cuff. A minute later I was in contact with the Woomera tracking station, which is about halfway across Australia. The Cap Com there informed me that the Woomera Airport lights were on, and he asked me if I could see them.

"Negative," I said. "There's too much cloud cover in this area. Sorry."

An hour and 13 minutes after launch I had left Australia behind and was in touch with the Canton Island tracking station, which is about half-way across the Pacific. I decided to have the first of two planned meals here. I pulled a squeeze-tube of apple sauce out of its receptacle and parked it out in the air in front of me. Weightless, it stayed put while I opened up the visor on my helmet. Then I squeezed the apple sauce into my mouth, and swallowed it without spilling a drop and closed up the visor again. There was no problem. I could see the brilliant blue horizon coming up behind me now; the sunrise was approaching.

The strangest sight of the entire flight came a few seconds later. I was watching the sunrise, which suddenly filled the scope with a brilliant red, and had put a filter onto the scope to cut down the glare. Then I glanced out of the window and looked back towards the dark western horizon. It was a startling sight. All around me, as far as I could see, were thousands and thousands of small, luminous particles. I thought for a minute that I must have drifted upside down and was looking up at a new field of stars. I checked my instruments to make sure that I was rightside up. Then I looked again. I was in contact with the Canton Island tracking station at the time, and I tried to tell the Cap Com there what it was like.

"This is Friendship Seven," I began. "I'll try to describe what I'm in here. I am in a big mass of very small particles that are brilliantly lit up like they're luminescent. I never saw anything like it. They're coming by the capsule, and they look like little stars. A whole shower of them coming by. They swirl around the capsule and go in front of the window and they're all brilliantly lighted. They probably average seven or eight feet apart, but I can see them all down below me, also."

The Canton Island Cap Com came on the air and asked if I could hear any impact between the particles and the capsule.

"Negative," I reported. "They're very slow; they're not going away from me more than maybe three or four miles per hour. They're going at the same speed I am approximately. They're only very slightly under my speed. They do have a different motion, though, from me because they swirl around the capsule and then depart back the way I am looking."

The particles seemed to disappear in the glare as soon as the sun came up. But I saw them again under the same conditions on the next orbit. This time, although I was having a few troubles with the capsule, I turned it around 180 degrees in order to look at the particles from another direction. I wanted to see if perhaps they were emanating from the capsule itself. They did not appear to be, however. They were not centered around the capsule but were stretched out as far as I could see. I saw fewer of them this time, because I was looking against the sun. But some of them still came drifting towards me, just as they had done when I first saw them. They were yellowish green in color, and they appeared to vary in size from a pinhead to perhaps three-eighths of an inch. They had the same color, luminous quality and approximate intensity of light as fireflies, and the sensation as I slowly rode through them was like walking backwards through a pasture where someone had waved a wand and made all of the fireflies stop right where they were and glow steadily.

I saw the particles once more on the third orbit, again just as the first rays of the sun appeared over the horizon. They stayed in sight for about four minutes, some of them turning dark as they went into the shadow of the capsule, others swirling up past the window and changing direction as I moved through them. It was a fascinating spectacle, and though various scientists have assumed since that the particles were undoubtedly emanating from the capsule itself, I found this hard to believe. I thought at first that they might be a layer of tiny needles that the Air Force had sent into space on a communications experiment and had then lost. But needles would not have been luminescent—nor was I at the proper altitude. I also thought that they might be tiny snowflakes formed by the condensation of water vapor from the control nozzles. I intentionally blipped the thrusters to see if they gave off particles. They gave off steam, but no particles that I could see. The particles were a mystery at the time, and they have remained one as far as I'm concerned. Our staff psychiatrist, Dr. George Ruff, heard me describe them at one of the debriefings after the flight, and he had only one question: "What did they say, John?" I guess they were as speechless as I was.

Though there was absolutely no connection between the two, I started having trouble with the automatic control system about fifteen minutes after I saw the particles for the first time, just as I was nearing the California coast on the first orbit. The capsule started to stray off to the right on the yaw axis. It would drift about 20 degrees; then the automatic pilot would sense the error and activate the large nozzles to swing it back into line again. The capsule kept cycling back and forth like this, between error and correction, until finally I switched off the ASCS altogether and started to control the capsule manually. There was no danger that this malfunction would move the capsule off course. We were on a predetermined path which had been set when we first went into orbit. But this constant repetition of error meant that the automatic system was using up hydrogen peroxide fuel at an excessive rate as it made each big correction. This would soon deplete the fuel supply unless we took precautions. I decided to remain on fly-by-wire, the system we use to manipulate the automatic thrusters with the manual control stick. There was no question at this point about going on. We were already committed to a second orbit before the control problem cropped up, and I was satisfied that I could keep things under control with the manual and fly-by-wire systems. The problem did mean, however, that I had to cut down on many of the other activities which I planned to carry out during the second and third orbits, for much of my time from now on was spent controlling the capsule. I had to cancel several of the experiments and observations which I wanted to make, including a series of tests of the sun's corona, some measurements of the brightness of the clouds, a second meal —a tube of mashed-up roast beef—and some further tests of a pilot's ability in space to adapt himself to darkness. I was also unable to take as many pictures as I had intended. I had intended putting the capsule on automatic controls while I concentrated on these other problems. I believe that we more than made up for the things that we had to leave out, however, by the fact that the troubles we had proved the validity of having a man in space. I was able to intercede and take over when the control system acted up. It is probable that the capsule would never have completed three orbits or might not have returned to earth at all if a man had not been aboard to exercise human judgment and control over the spacecraft machine.

John Glenn did not know it at the time, but Mercury Control had picked up a much more serious problem as the capsule passed over the Canaveral area and headed for its second orbit. The Control Center received a telemetry signal from the capsule which indicated that the ablative heatshield on the blunt nose of the capsule had come loose. The shield is designed to come loose during the final stage of descent towards the ocean so it will extend a perforated skirt, designed to hang beneath the capsule as part of an impact bag that takes up some of the shock of landing. The shield must remain locked to the capsule during re-entry, for it is the only means of protecting the Astronaut from the tremendous heat which builds up outside as the capsule penetrates the atmosphere. John Glenn did not know at this time that the people on the ground were concerned. One of two switches mounted on the base of the capsule had sensed the status of the locks which hold the shield in place and their signals going directly to the ground had warned that the shield was deployed. John did not have the same indicator in his capsule. This caused a good deal of concern in the Control Center, to say the least. Telephone calls were put through immediately over open circuits to the McDonnell factory in St. Louis, and engineers pored over all of their wiring diagrams and specifications trying to determine what might be wrong. They knew it was possible that the switch itself was faulty and that it had sent an erroneous signal. It was also possible, however, that the signal was correct. If this was true, John Glenn was in for an uncomfortable and perhaps tragic ride to earth. There was one possible way out, however. Heavy metal straps which held the package of retro-rockets in place under the heatshield were attached directly to the capsule. So long as they stayed in place, they might also help hold the shield against the capsule. Normally, the retro-package would be jettisoned immediately after the rockets had been fired, and the straps would go with it. This would be done so the heatshield would be clean and smooth and would ablate away up the re-entry heat most efficiently. Leaving the retro-package in place would spoil this arrangement. But it might save the shield—and John Glenn. The Control Center decided that this was the safest course to follow. It also decided not to burden John Glenn with such a tremendous cause for worry until it had been able to check on the problem in greater detail. The tracking stations were informed, however, and were asked to monitor the situation and to ask John Glenn a few calm questions about the status of his heatshield as he entered their respective areas.

During much of the second orbit I worked over the control system, trying to pin down a pattern of errors so I could determine what was wrong and make allowances for it. I could hear and feel the large thrusters outside of the capsule as they popped off their bursts of hydrogen peroxide, first in one direction and then in the other. I could feel the slight throb of the smaller nozzles when I cut them in. As I was crossing the Atlantic the second time, the problem I had been having seemed to reverse itself, and I reported this to the Cap Com on board the Atlantic ship tracking station.

"At one time," I said, "I had no left low thrust in yaw; now that one is working and I now have no low right thrust in yaw. Over."

I confirmed this report over Kano, Nigeria. All of this routine kept me pretty busy. I had to keep my hands busy with the controls most of the time, but I thoroughly enjoyed it. The idea that I was flying this thing myself and proving on our first orbital flight that a man's capabilities are needed in space was one of the high spots of the day.

The problems had increased by the time I reached the Indian Ocean the second time. Something had gone wrong now with the ASCS indicators, and the various attitudes of yaw, pitch and roll which the instruments presented did not jibe at all with what I could see just by looking out of the window. The Cap Com on the Indian Ocean ship asked me if I had noticed any constellations yet. I answered that I was too busy paying attention to the control system to identify any stars. He also gave me my first clue that something might be wrong with the heatshield.

"We have message from MCC" (Mercury Control Center), he said, "for you to keep your landing bag switch in OFF position, landing bag switch in OFF position. Over."

I rogered for the transmission, and I thought right away that the ground must be getting some peculiar indications. I was to get a better inkling of what was going on about seven minutes later when I checked in with Gordon Cooper in Australia.

"Will you confirm the landing bag switch is in the OFF position? Over," said Cooper.

"That is affirmative," I said. "Landing bag switch is in the center OFF position."

"You haven't had any banging noises or anything of this type at higher rates?" Cooper asked.

"Negative," I said.

It was clear to me now that the people down on the ground were really concerned or they would not be asking such leading questions. I was fairly certain, however, that everything was in good shape. It occurred to me that if the heatshield were really loose, I would almost certainly have been able to hear or feel it shaking behind me or banging against the edge of the capsule as we drifted back and forth on those large deviations in yaw. I had heard nothing. Still, there was room for concern. As we have explained earlier in the book, the heatshield is made up of a thick coating of resinous material which is designed to dissipate most of the heat and energy picked up during re-entry and get it out of the capsule's system by melting and boiling away very slowly. This was the only thing that stood between me and disaster as we came through the atmosphere. If it was not tightly in place, we could be in real trouble.

I was having a few other minor problems along about this time. I had two warning lights shining on the panel. One showed an excess of water in the cabin environmental control system. This could mean that not all of the water was turning into steam and popping off outside through the outlet valve as it was supposed to. The danger was that the water which did not bleed off as steam could freeze up the outlet valve and clog the system. I turned down the volume of water which was running through the system, and this had the proper effect. The light would come on again every time I tried to increase the water flow substantially. The system never did clog up, however. The other warning light showed that we had used up more fuel than we should have at this point in the automatic control system. Considering all of the erratic control maneuvers which the system had been making, this was not surprising. We still had plenty of fuel left, however. The light is designed to flash on when you have 65 percent of your supply remaining in the tank, simply to make you pay attention to the problem. Another problem was that the supply of oxygen in the secondary tank started to decline even though I was not using it. The people on the ground picked this up by telemetry and asked me at one point how much of this supply I had been using. I told them I had not touched it. There must have been a leak somewhere in the system, but fortunately I did not need to draw on this supply, so it was not a serious matter. It could have been, however, and the technicians went over the system with a fine-tooth comb when they got the capsule back. It was a fault we would want to repair before any future flights.

As I passed over Canton Island on the second orbit I saw the particles again. I tried to photograph them, but apparently there was not enough light for the color film and none of the pictures came out. The heatshield problem came up again as I was putting the camera away. The Cap Com at Canton Island put it this way: "We also have no indication that your landing bag might be deployed. Over."

I asked if someone else had reported that it could be down, and he said, "Negative. We had a request to monitor this and to ask you if you heard any flapping."

"Negative," I said. I was still not overly concerned because I had had no indication in the capsule that anything was wrong. Looking back on the whole event, I realize that the controllers were trying to keep me from being worried about the situation. I really don't think, however, that you ought to keep the pilot in the dark, especially if you believe he might be in real trouble. It is the pilot's job to be as ready for emergencies as anyone else, if not more so. And he can hardly be fully prepared if he is not being kept fully informed. On future space flights, when the spacecraft and its crew will get thousands of miles from earth, some of the apron strings will have to be cut. On this flight, of course, it was the first time that the man had been that far away from home, and the entire family was naturally concerned.

I did have some doubts along about this time as to whether we would go for a third orbit. The automatic controls were misbehaving; the manual controls that I relied on most had become a little mushy—at least, they were not as crisp as they had been. With the problems we were having, I was concerned that perhaps the people down on the ground might prefer for me to come on home. I sincerely hoped not. There was nothing to be concerned about unless the problem got significantly worse. We were still in good shape and I felt that if it was going to be necessary for me to bring the capsule back myself, I might as well have another 90 minutes of practice at the controls. The people on the ground apparently felt the same way. I was very happy when Mercury Control recommended that I go for the third orbit. We talked about it as I passed over Hawaii, and I concurred 100 percent. As I crossed California, Wally Schirra gave me the temperatures on the inverters which the ground stations had picked up by telemetry. The inverter on the fans was 215 degrees and the ASCS inverter was 198 degrees. These were both a little high, but Wally said they recommended that I not do anything about it. We had discovered in recent tests that the inverter could stand more heat than we had originally anticipated. "It looks real good," Wally said. He also informed me that my elapsed-time clock was running about a second slow. Since this would effect the sequence time for firing the retros, he gave me a new reading and told me to subtract one second from the other readings that I had already jotted down in order to compensate for the error in the clock. We would actually reset the clock in another four or five minutes, he said, when I checked in with Al Shepard over Cape Canaveral.

Al then told me to reset the clock manually so that it read 04:32:38. This meant that the retro-rockets would start firing 4 hours, 32 minutes and 38 seconds after launch. This would be in another hour and 24 minutes. As I was talking to Al I could look down and see the entire state of Florida and clear back to the delta of the Mississippi River around New Orleans. This was the best view I had had of the U.S. There was a cloud deck to the north, but I could see as far north as North Carolina. To the south I spotted islands east of Cuba. I looked out over the Atlantic to check the recovery area where I would be landing the next time around.

There were a few scattered clouds but no sign of a major weather system. The sea seemed to be placid, though it was so far below me I could not really tell. I had noticed the wake of a ship on a previous pass over the Atlantic, and I assumed it might have been one of the three aircraft carriers, standing by waiting to launch helicopers to pick me up.

Gus Grissom and I had a brief chat as I passed over his Bermuda station for the last time. I told him about the view I had and that it looked good in the recovery area. "Very good," Gus said. "We'll see you in Grand Turk."

"Yes, sir," I answered. This was the last lap and I felt that we had things well under control. Gus relayed a message from the Cape that they recommended I use the automatic control system during re-entry and back it up with the manual controls. I pointed out that the ASCS had been very erratic and that I had not been able to pin it down to any particular item. It had gone wrong in pitch, yaw and roll. I was going to reserve that decision until closer to retro-fire time. Gus said they had recorded this information and asked me to read off the oxygen supply. I still had 62 percent left in the primary tank and 94 percent left in the secondary tank. We had lost 6 percent from the secondary tank without even tapping it, but this was no problem since I had not needed it anyway.

The Cap Com asked me, as I passed over the Canary Island station for the last time, if I was still seeing those particles I had talked about. Apparently everyone was fascinated by this phenomenon. I told him that I had seen a few just after I left Canaveral, and that I knew they were not coming from the capsule because they were moving towards me. The aeromed surgeon at the Canary Station came on to ask me again if I had experienced any nausea. I told him that I had not, and that I had felt perfectly normal during the entire flight. "I feel fine," I said.

A few seconds later the Atlantic Ocean tracking ship contacted me. I gave them another reading on fuel, oxygen and amps. We were down to 64 percent on the manual control system now and to 54 percent on the automatic system. The gauge showed that we still had 62 percent of the oxygen left in the primary tank and 94 percent in the secondary. The amperes stood at 23. Then I gave the ship a long message describing the status of the attitude control system and asked the Cap Com on the ship to pass it along to Cape Canaveral. I told him that I had let the capsule drift 180 degrees off center, in order to see what would happen, and was now trying to reorient it again.

"When I am all lined up with the horizon and the periscope," I explained, "my attitude indications on the instrument panel are way off. My roll indicates thirty degrees right; my yaw indicates thirty-five degrees right; and pitch indicates plus forty degrees. I repeat, plus forty when I am in orbit attitude." Roll and yaw should have been zero and pitch should have read minus 34 degrees. This meant that the gyros in the autopilot were not keeping up with the actual attitude of the capsule and were giving me misleading information. The situation was not serious at the moment. So long as I could see the horizon outside and line the capsule up against it by myself, I knew that I could keep it properly aligned. It did mean, however, that I could not trust the ASCS to control the capsule while I attended to other matters. I would have to control it accurately myself or we might not be in the correct attitude when it came time to fire the retros. The capsule's attitude would have to be near-perfect when the rockets fired, or the angle of re-entry would be affected. If we were too far off at that moment, I might have trouble getting down.

I turned the capsule to the left a little on the yaw axis as I crossed the Atlantic in order to get a better view of my third sunset of the day. I described the sight to the Atlantic Ocean tracking ship and said that I could see a very orange band along the horizon, then a lighter yellow on top of that, followed by a very deep blue, a very light blue and then the black sky. I also reported that it was not too easy to see anything through the window when I was looking towards the sun. It appeared, I said, as if "we might have smashed some bugs on the way up off the pad. Looks like blood on the outside of the window." There was no attendant around to wash this windshield.

As I passed the tip of Africa and started over the Indian Ocean I could see a huge storm front stretching out beneath me as far as I could see. It was dark now, and the ocean itself was covered with a thick layer of clouds. But I could see bright flashes of lightning inside the clouds. The weather people had wondered whether I would be able to see lightning from such a high altitude. The flashes showed up brilliantly, like flash-bulbs being popped off behind a white sheet. Each flash lit up an entire bank of clouds. I reported this to the Zanzibar tracking station as I passed near it for the last time. I did not see any of Africa on this pass; the third orbit took me further south than I had gone before and the entire area was dark and cloudy.

Three hours, 59 minutes and 15 seconds after launch I was in contact once more with Gordon Cooper at the Muchea station in Australia. We chatted briefly about the troubles I was having—the ASCS was still not functioning properly. But I told him I was still able to correct the errors with the manual system, and I felt in fine shape. I asked him to send a message to General David Shoup, the commandant of the U.S. Marine Corps.

"This is Friendship Seven," I told Gordon over the radio. "In forty-five seconds I would like to have you send a message for me, please. I want you to send a message to the commandant, U.S. Marine Corps, Washington. Tell him I have my four hours required flight time in for the month and request flight chit be established for me. Over."

"Roger," Cooper answered. "Will do. Think they'll pay it?"

"I don't know," I said. "Gonna find out."

"Roger," said Cooper. "Is this flying time or rocket time?"

"Lighter than air," I answered.

Gordon and I then discussed the various readings on my instrument panel and the positions of all the switches. The temperature inside the suit at this point was 70 degrees Fahrenheit. The cabin temperature was 90 degrees. The pressure inside the suit was 5.8 pounds per square inch. The cabin pressure was holding at 5.5 psi. The amps stood at 24. I had 60 percent of the oxygen left in the main tank, 90 percent in the secondary tank—which meant that it was still leaking. The relative humidity in the cabin was 36 percent. The temperature on the ASCS inverter was 115 degrees. The fan inverter was 110 degrees. All of the warning lights were out except the fuel quantity light, and I had turned off the audio warning signal on this light to cut down on the noise level. I told Cooper that if the ASCS did not respond I would stay on manual control throughout the retro sequence.

I started to stow the loose equipment away as I checked in with the Canton Island station for the last time, so that nothing would get in my way during the re-entry phase. A few minutes later I was over the Hawaii tracking station and gave the gauge readings to the Cap Com there. I had 43 percent fuel remaining in the automatic system, 45 percent in the manual system. The oxygen supply was still 60 percent and 90 percent. The amps stood at 23. Then the Cap Com asked me another leading question about the heatshield.

"Friendship Seven," he said, "we have been reading an indication on the ground of segment 5-1, which is landing-bag deploy. We suspect this is an erroneous signal. However, Cape would like you to check this by putting the landing-bag switch in auto position and see if you get a light.

Do you concur with this? Over." I thought that one over for a few seconds.

"O K.," I answered, "If that's what they recommend, we'll go ahead and try it. Are you ready for it now?"

"Yes," the Cap Com said, "when you're ready."

This was a rather tricky thing to do, and I was a little reluctant to try it. The idea was that if I turned my landing-bag deploy switch to the automatic position and the light on my instrument panel turned green, this would indicate that the shield had indeed deployed on its own and that it was loose. At least, we would know. I was slightly concerned about the reverse side of this coin, however. What if the shield had not deployed but, because the system was malfunctioning, would decide to do just that when I switched it to automatic? Then we would have jumped from the frying pan into the fire. I knew that personnel at the Cape would never make such a recommendation without considering all the aspects of the problem carefully, so I went ahead and rapidly turned the switch on and off again. The light did not come on. This was a pretty good indication to me that we were in good shape. The Hawaii Cap Com seemed to think so, too, for he rogered for my message and said, "That's fine. In this case, we'll go ahead, and the re-entry sequence will be normal."

As it turned out, he was a bit premature. I received different instructions a little later on, and the re-entry sequence was not normal at all. The Hawaii Cap Com also instructed me to change the retro clock again by one second. Apparently we had gone just a wee bit faster than our original predictions, and we lopped one more second off the elapsed time to make it 4 hours, 32 minutes and 37 seconds. This was a total change of 2 seconds on the clock during the entire trip, which was not bad at all on an orbital mission which lasted 16,357 seconds from launch to retro-fire. The flight surgeon at the Hawaii station asked me if I was still comfortable. I told him that I was in very good shape. A light on my instrument panel went on to signal that I had exactly 5 minutes left before retro-fire, and the Hawaii Cap Com started to give me a time hack based on Greenwich Mean Time so that I could double-check the schedule. Our communications faded out, however, just before we could complete this important transmission, and I was a little uncertain about exactly when the retro-rockets should be fired. From what I thought I had heard, there seemed to be a small discrepancy between my time and the time they had computed on the ground, and this bothered me. The timing had to be precise, since at my orbital speed of 5 miles per second an error of a single second would mean a dispersion of 5 miles in the impact area. I tried to call Wally Schirra, who was standing by at the Point Arguello station in California, but it took about a minute and a half before we had good communication. It seemed quite a bit longer at the time. By the time I finally got through to Wally and explained the timing discrepancy to him, I had only 50 seconds left before the retro sequence was due to start. Wally worked fast and confirmed the timing. We had 45 seconds to go, he said. I told him that I was on ASCS, which was working well at this point, and backing it up with the manual system. Wally rogered for this message and said that I had 30 seconds to go before the retro sequence began. I rogered for this and told him that my 30-second retro-warning light was on. Then, for the next couple of minutes, while I went through the most critical phase of the entire flight, Wally and I kept up a fairly constant exchange of messages. They went like this:

SCHIRRA: John, leave your retro-pack on through your pass over Texas. Do you read?
GLENN: Roger.
SCHIRRA: Fifteen seconds to sequence.
GLENN: Roger.
SCHIRRA: Ten.
SCHIRRA: Five, four, three, two, one, mark.
GLENN: Roger. Retro sequence is green. (*There was a 30-second built-in delay here.*)
SCHIRRA: You have a green. You look good on attitude.
GLENN: Retro attitude is green.
SCHIRRA: Just past twenty.
GLENN: Say again.
SCHIRRA: Seconds.
GLENN: Roger.
SCHIRRA: Five, four, three, two, one, fire.
GLENN: Roger, retros are firing.
SCHIRRA: Sure, they be.
GLENN: Are they ever. It feels like I'm going back towards Hawaii.
SCHIRRA: Don't do that. You want to go to the East Coast.
GLENN: Roger. Fire retro-light is green.

SCHIRRA: All three here. (*Meaning that Schirra could see telemetry indications that all three rockets had fired on schedule.*)

GLENN: Roger. Retros have stopped.

SCHIRRA: Keep your retro-pack on until you pass Texas.

GLENN: That's affirmative.

SCHIRRA: Check.

SCHIRRA: Pretty good flight from all we've seen.

GLENN: Roger. Everything went pretty good except for all this ASCS problem.

SCHIRRA: It looked like your attitude held pretty well. Did you have to back it up at all?

GLENN: Oh, yes, quite a bit. Yeah, I had a lot of trouble with it. (*He was referring here to the general behavior of the ASCS during the flight and not just to the retro-fire sequence when the ASCS was working well.*)

SCHIRRA: Good enough for government work from down here.

GLENN: Yes, sir, it looks good, Wally. We'll see you back East.

SCHIRRA: Rog.

GLENN: All right, boy.

GLENN: Fire retro is green.

SCHIRRA: Roger.

GLENN: Jettison retro is red. I'm holding onto it.

SCHIRRA: Good head.

GLENN: I'll tell you, there is no doubt about it when the retros fire.

The three retro-rockets had fired on schedule at five-second intervals. Each one gave me a very solid push, and since I was weightless at the time and they were firing forward, against the direction of the flight, I had the distinct sensation of accelerating back in the direction I had come from. Actually, the rockets were only slowing me down by about 500 feet per second. Both Al Shepard and Gus Grissom had experienced the same sensation when they tested out the retros on their flights, and I was prepared for it. The firing of the rockets caused some motion of the capsule, but since I was using both the automatic and manual control systems together, I maintained the proper attitude during retro-firing and then realigned Friendship 7 for its descent through the atmosphere.

During this phase of the flight, John Glenn's heart was beating at the rate of 96 beats per minute. It had averaged 86 beats per minute during the preceding three hours of flight. As the re-entry phase continued, his pulse increased to 109 beats per minute, and it reached a peak of 134 beats per minute just before the drogue parachute deployed on the final descent and the capsule was going through a period of high oscillations.

Approximately 5 minutes after the retro-rockets fired, I came into radio contact with the tracking station in Texas. I had rather hoped that by this time the people on the ground could tell me to go ahead and jettison the retro-pack. We did not know exactly what affect the retro-pack might have on the even distribution of heat over the shield. It was just possible that it might cause hot spots to break out and damage the shield before we had completed re-entry. Holding onto the pack would also upset the normal chain of events which the capsule was supposed to perform automatically. The wiring was so arranged, for example, that if the retro-pack did not drop, the circuits which were normally set in motion by this event would fail to retract the periscope automatically and close the door behind it to keep out the heat. They would also not be in a position to respond to a special switch which is thrown automatically when the G forces of re-entry reach .05 G, at which time other automatic sequences start for various events involved in the landing maneuver. All of this was supposed to be done automatically, but, for safety reasons, it all hinged on the jettisoning of the retro-pack. If the pack was not jettisoned automatically, I would activate these other functions as well. I was prepared to do all of this, of course, but it was not the way we had planned it; and since any deviation from the standard procedure always leaves a certain amount of room for doubt and suspense, I frankly hoped that we could make a normal descent. I was somewhat concerned, therefore, when the Texas Cap Com sent me the following transmission 17 minutes before I was due to land:

"This is Texas Cap Com, Friendship Seven," he said. "We are recommending that you leave the retro-package on through the entire re-entry. This means that you will have to override the .05 G switch which is expected to occur at 04:43:53. This also means that you will have to manually retract the scope. Do you read?"

"This is Friendship Seven," I said. "What is the reason for this? Do you have any reason? Over."

"Not at this time," the Cap Com answered. "This is the judgment of Cape Flight. . . . Cape Flight will give you the reasons for this action when you are in view."

Some 30 seconds later I came within range of the Control Center at the Cape and heard Al Shepard's voice over the radio.

"Recommend that you go to re-entry attitude and retract the scope manually at this time," Al said.

"Roger," I said. "Retracting scope manually." I reached down for the handle and began to pump the scope in. It came all the way, and the door closed tightly behind it.

"While you are doing that," Al said—and here I finally learned for certain what the problem was—"we are not sure whether or not your landing bag has deployed. We feel it is possible to re-enter with the retro-package on. We see no difficulty at this time in that type of re-entry. Over."

"Roger," I said. "Understand."

"Estimating .05 G at 04:44," Al added.

It was now 04:41, or 4 hours and 41 minutes after launch. In another few minutes, I would be in the middle of the hottest part of my ride. The automatic control system had been acting up again—drifting off center and then kicking itself back into line again—so I was now controlling almost completely by the manual stick. The fuel in the manual system was running low, however—the gauge read that I had only about 15 percent left in the tank. So I switched to fly-by-wire in order to draw on what fuel was left in the automatic system. I was still controlling manually, however, since this was the advantage that the fly-by-wire system provided. I used the manual stick; the nozzles that I activated and the fuel that I expended belonged to the automatic system.

During the final descent through the atmosphere, the blunt nose of the capsule had to be kept pointed so that the heatshield would hit the particles of the atmosphere first. If the capsule were not properly aligned, some of the intense heat could spill over the edge of the shield and flow back along the sides of the capsule, which are not nearly so well protected against the extreme temperatures building up during re-entry. In addition, the capsule might start to oscillate quite a bit as it was buffeted by the atmosphere. That is, it would sway back and forth. This could also let some of the heat impinge on the side of the capsule. As I have said before, it was not the kind of re-entry that we had hoped for. But we had included all of these eventualities in our training, and I was set. It was going to be an interesting few minutes, however.

As we started to heat up on re-entry, I could feel something let go on the blunt end of the capsule behind me. There was a considerable thump, and I felt sure it was the retro-pack breaking away. I made a transmission to Al Shepard to this effect, but he apparently did not hear me. By this time, the capsule was so hot that a barrier of ionization had built up around it and cut off all communications between me and the people on the ground. This was normal, and I had expected it to happen, but it left me more or less alone with my little problem.

Just 24 seconds before John Glenn made this transmission, Al Shepard had started to recommend to Glenn that he jettison the retro-pack as soon as the Gs built up to 1 or 1.5. Glenn did not receive the message, however. The communications black-out had already set in.

I saw one of the three metal straps that hold the retro-pack in place start to flap around loose in front of the window. Then I began to see a bright orange glow building up around the capsule. "A real fireball outside," I said into the microphone. The loose strap burned off at this point and dropped away. Just at that moment I could see big flaming chunks go flying by the window. Some of them were as big as 6 to 8 inches across. I could hear them bump against the capsule behind me before they took off, and I thought that the heatshield might be tearing apart. As it turned out later, these were parts of the retro-pack breaking up. It had not fallen away after all, and the heatshield itself was coming through in perfect shape. This was a bad moment. But I knew that if the worst was really happening it would all be over shortly and there was nothing I could do about it. So I kept on with what I had been doing —trying to keep the capsule under control—and sweated it out.

I knew that if the shield was falling apart, I would feel the heat pulse first at my back, and I waited for it. I kept on controlling the capsule. It was programed to do a slow, steady spin on its roll axis at the rate of 10 degrees per second. The purpose of this maneuver was to equalize the aerodynamic flow around the capsule and to keep it from exceeding the limits that we had estimated were maximum for re-entry. The automatic control system was normally supposed to handle this procedure but I kept control with the manual stick and did it myself. Pieces of flaming material were still flying past the window during this period, and the glow outside was still bright and orange. It lasted for only about a minute, but those few moments ticked off inside the capsule like days on a calendar. I still waited for the heat, and I made several attempts to contact the Control Center and keep them informed.

"Hello, Cape. Friendship Seven. Over. Hello, Cape, Friendship Seven. How do you receive? Over." There was no answer.

Down in the Control Center at this point, the men at the consoles were definitely worried. They were still tracking the capsule's descent on radar and they knew where it was and that it still seemed to be intact. But they were deeply concerned over the fate of the heatshield—and of John Glenn. They knew from previous tests that the temperature of the shield would be about 3,000 degrees Fahrenheit. The temperature of the heat pulse which had built up around the capsule would stand at about 9,500 degrees Fahrenheit—or slightly less than the temperature of the sun itself. The combination of this concern plus the absolute silence in their headsets was almost unbearable. The communications black-out lasted for 4 minutes and 20 seconds. Here, too, the seconds passed "like days on a calendar." Someone behind Commander Shepard's console said, "Keep talking, Al." Shepard spoke once more into the microphone anchored in front of his lips.

"Seven, this is Cape," he said. "How do you read? Over."

This time John Glenn heard the transmission. It was 4 hours, 47 minutes and 11 seconds after launch, with 7 minutes and 45 seconds to go before Glenn's capsule was to hit the water.

"Loud and clear," I said. "How me?"

Al's voice really sounded welcome when it finally came through.

"Roger," Al said, cool as ever. "Reading you loud and clear. How are you doing?"

"Oh, pretty good," I said.

The heat had never come. Instead, the high temperature pulse began to simmer down and the glow gradually disappeared. The Gs built up to a peak of about 8 now, but they were no problem. Al informed me that they had worked out my impact point in the recovery area and that I should be landing within one mile of one of the destroyers.

"My condition is good," I said, "but that was a real fireball, boy."

Twelve seconds after this transmission I reported that the altimeter read 80,000 feet. Nineteen seconds later we were at 55,000 feet. The capsule was rocking and swaying quite a bit at this point, and I was having trouble controlling it. We were almost completely out of fuel in both control systems by now. But, even if I had had sufficient fuel, we were now so far into the thick atmosphere that the control nozzles would not have had much effect on the capsule's movement. I decided to deploy the small drogue chute a few moments before it was due to come out, and damp the oscillations that way. The capsule beat me to it, however. I was just reaching for the switch to override the automatic timer when the drogue chute broke out on its own. I could feel the thud of the mortar which launched it. The window was covered now with a thin layer of melted resin that had streamed back from the heatshield. However, I could still see the drogue open up at 30,000 feet. This was about 9,000 feet higher than where we would normally break it out. The chute held, and the capsule began to settle down into a much smoother descent. The swaying was cut sharply. I had to pump the periscope out by hand since we had interrupted the automatic sequence. At about 20,000 feet, the snorkels opened up to let in outside air. At 10,600 feet, a barometric switch started the landing sequence. Through the periscope and the coated window, I saw a marvelous chain reaction set in. I watched the antenna canister—which housed the chute—detach. It dragged the main chute along behind it, still wrapped up inside its bag. When the shrouds of the chute had stretched out to their full length, the bag peeled off and left the chute, still in a reefed condition, trailing out like a long ribbon straight above me. Then, when the chute was partially full of air and had found its proper position, the reefing lines broke away and the huge orange and white canopy blossomed out, pulsed several times and was steady. I could feel the jolt in the cabin as we slowed. From the indications of the instruments we seemed to be dropping a few feet per second faster than I thought we should. But I studied the chute closely through the periscope and window, and it appeared to be in such perfect shape, with no rips or holes in it, that I decided not to use the reserve chute which was still packed away in the roof of the capsule and available if I needed it. It was a moment of solid satisfaction. As I told Al over the radio—with a real trace of relief and some excitement in my voice, I guess—it was a "beautiful chute." It was a wonderful sight to see that good chute open up.

I was descending now at the rate of 42 feet per second and had 5 minutes and 10 seconds left before impact. I contacted the destroyer *Noa*, which had the code name of "Steelhead," and told the skipper that my condition was good but that it was a little hot inside the capsule. He informed me that he had picked up on his radar the chaff which the main chute had kicked out and that he was heading in my direction. He estimated that it would take him about an hour to get on station.

I started to run down the checklist of landing procedures. I unfastened from my pants leg the plug which connected the biomedical sensors. I removed the blood-pressure equipment from the suit, loosened the chest strap and got it free, unhooked the respiration sensor from my lip mike and stuffed it inside my suit, disconnected the oxygen exhaust hose from the helmet and unstowed the survival pack that I had to the left of my couch and kept it handy in case of an emergency. Al Shepard got on the radio at this point to make sure that my landing-bag light was on green so that it would deploy and take up the shock of landing.

"That's affirmative," I said. "Landing bag is on green."

Then Al came on again to recommend that I remain in the capsule unless I had "an overriding reason for getting out." He knew that the destroyer was only about 6 miles from where I would land and that instead of using helicopters to pick me up as we had planned, I would have to be hoisted aboard by the destroyer. It would be simpler in this case if I stayed shut up inside so we would not take any chance on losing the capsule. We had rehearsed this method of recovery as well as the helicopter method, so I was prepared for either one and I rogered for his message. I kept up a running account now of my approach to the water so that everyone on the network would know my status.

"Friendship Seven," I said, 48 seconds before I hit. "Ready for impact; almost down."

Fifteen seconds later: "Friendship Seven. Getting close. Standing by."

Twenty seconds later: "Here we go."

Ten seconds after that: "Friendship Seven. Impact. Rescue Aids is manual."

I pushed the button which started the flashing light on top of the capsule and the automatic radio signals which would help the recovery force home in on my position.

The capsule hit the ocean with a good solid bump, and went far enough under water to submerge both the periscope and the window. I could hear gurgling sounds almost immediately. After it listed over to the right and then to the left, the capsule righted itself and I could find no traces of any leaks. I undid the seat strap now and the shoulder harness, disconnected my helmet and put up my neck dam so I could not get water inside my suit if I had to get into the ocean. I was sweating profusely and was very uncomfortable. I kept the suit fans going, but they did not help much. The snorkels in the capsule wall were pumping in outside air, but it was extremely humid outside and this did not help to cool me off one bit, either. I thought about removing the lid of the capsule and climbing on out. But I decided against it. I knew that any body movement would only generate more heat and make me even warmer. The thing to do was sit tight, stay motionless and try to keep as cool as possible.

"Steelhead" kept up a running commentary on how she was doing. First, she was 4 minutes away, then she slowed down and was 3 minutes away; then her engines were stopped and she was coming alongside. The capsule window was so clogged now with both resin and sea water that I could not see her. Strangely enough, however, the capsule bobbed around in the water until the periscope was pointing directly at the destroyer, and it kept her in view from then on. I could read her number—841—and I could see so many sailors in white uniforms standing on the deck that I asked the captain if he had anybody down below running the ship. He assured me he did. Then he drifted alongside very slowly until we gently bumped into each other.

Two sailors reached over with a shepherd's hook to snag the capsule, and moments later we were on deck. I started to crawl through the top to avoid blowing the side hatch and jiggling the instruments inside the capsule. I was still so uncomfortably hot, however, that I decided there was no point in going out the hard way. After warning the deck crew to stand clear, and receiving clearance that all of the men were out of the way, I hit the handle which blew the hatch. I got my only wound of the day doing it—two skinned knuckles on my right hand where the plunger snapped back into place after I reached back to hit it. Then I climbed out on deck. I was back with people again.

I was still very hot inside the space suit, and the first thing I wanted to do was get it off. We went to the captain's cabin, where the crew helped me slip out of the sleeves. That felt pretty good. I peeled off the rest of the suit and felt still better. Then, in a pair of long-handled space skivvies which were wringing wet with perspiration, I stepped out on the deck and stood in the breeze. That felt best of all. I showered after that, and drank a large glass of iced tea. The doctors aboard the *Noa* gave me a physical examination. They were not equipped to take an electrocardio-gram or chest X-rays, and the ship was rolling too much to weigh me accurately, so we left these items until I could get to the carrier.

President Kennedy called the *Noa*, and we talked for a couple of min-utes. He was very kind and expressed great interest in the flight and how I was feeling. A little later I heard from Annie. I could tell by her voice that she was thrilled and relieved. Then I spent a short time out on deck debriefing myself into a tape recorder while the events were still fresh in my mind. NASA had sent kits ahead to the recovery ships with personal equipment that I might need until I could get back to Grand Turk and catch up with my own things. The kit included underwear, dark glasses, a wrist watch, a flying suit, a pair of sneakers and a blank check. I assumed that the check was to be made out to my own bank account to provide me with some spending money until I could get home. I had already thought of that, however, and had given Scott Carpenter some money to hold for me until we caught up with each other at Grand Turk. I did not fill out the check.

The men aboard the *Noa* were so pleased that it was their ship and not the carrier which had picked me up that they made me an honorary member of the crew and presented me with a fifteen-dollar check as the February winner of their "Sailor-of-the-Month" contest. I endorsed the check over to the ship's welfare fund. The chaplain on the destroyer presented me with a Bible, which he said was a "navigation chart to the heavens which I had just visited." It was a touching ceremony, and I told the men there was no other ship's crew I would rather belong to. I found out later that to commemorate the occasion the crew painted some white footprints on the deck where my boots hit when I jumped out. It was this kind of *esprit* and interest in the Navy, and in the other services which have cooperated with Project Mercury, that has made our work easier and very satisfying.

Just before sundown a helicopter arrived from the carrier *Randolph* to transfer me to that ship. I wanted to see the capsule once more before I left so I went back on the deck where it was tied in place and retrieved the ditty bag that was still lying inside it, which contained instruments, camera and film. The sun was setting now, and I took a good look. It was the fourth beautiful sunset of the day, and you don't often have days like that.

The helicopter had lowered a hoist and I was winched up inside it for the ride to the carrier. The carrier's crew was lined up on the hangar deck, cheering and applauding as I walked through. It was a very impressive and heart-warming experience. I completed the physical exam, had a bite to eat and then got ready to take off in a Navy S2F jet for the one-hour flight to Grand Turk. The pilot, Commander Hal Hamburger, asked me if I wanted to fly it. I said it had been so long since I had piloted an S2F that I had better not. I acted as copilot for Hal, however, and read off the checklist before we took off. We were launched off the deck. There was a bright moon overhead—my fourth moon of the day—and it was a good flight.

There was quite a welcoming committee waiting for me on Grand
Turk Island. It included two of the other Astronauts—Deke Slayton and
Scott Carpenter. Al arrived the next day. Gus Grissom joined us from
Bermuda and Wally Schirra got in a little later from California. There
were several NASA officials—Walt Williams and Bob Gilruth along with
a platoon of doctors and other experts waiting to debrief me. Under their
direction, I underwent another complete physical examination to determine what effects, if any, the flight had had on me from a physiological
standpoint. For one of the balance tests, I walked along a series of rails
laid out across the floor. Each one of them was narrower than the one
before it, and the purpose of this exercise was to see if the hours I had
just spent in a state of weightlessness had affected my sense of balance. I
did not fall off, so the doctors were satisfied. (Colonel Chuck Yeager, the
famous test pilot who broke the sound barrier a few years ago in the X-1,
told me later that this test only proved I had been brought up near some
railroad tracks.) Bill Douglas and his cohorts had another go at me, and
I had the usual number of pokings and probings and measurements that
we have all become used to now in this program.

*The examination on Grand Turk showed that John Glenn's blood
pressure was 128 over 78. (It had been 118 over 80 before launch.) His
pulse was now 72 beats per minute. (It had been 68 before launch and 76
on board the* Noa.) *His respiration was 14 breaths per minute, the same
as it was before the launch. His oral temperature was 98 degrees Fahrenheit (two-tenths of a degree lower than it had been before launch, but
using a different thermometer). The internist who looked him over found
no change in either his heart or his lungs. The X-rays showed no change
or abnormality. His weight had dropped 5 5/16 pounds between launch
and the examination aboard the carrier. This was approximately the same
loss in weight that he had experienced during simulations on the centrifuge, however, and the doctors attributed most of it to simple dehydration. Glenn had perspired a good deal from the time he landed until he
was picked up. And between the time he was launched and weighed on
board the carrier he had consumed only 724 cubic centimeters of fluids.
This included 119.5 grams of apple sauce, which was 78.7 percent water;
265 c.c. of iced tea aboard the* Noa; *240 c.c. of water; and 125 c.c. of
coffee. In comparison with this intake of fluids, Glenn had emitted 800
c.c. of urine during the flight, which the doctors described clinically as
"clear, straw-colored urine with a specific gravity of 1.016, pH 6.0, and*

negative microscopically and for blood, protein, glucose and acetone. This volume of urine was passed," the report continued, "just prior to retro sequence."

Glenn's eyes, ears, nose and throat were normal after the flight, as they were before it. Examination of his lower extremities showed "no swelling or evidence of thrombosis." Despite repeated movements of his head during the flight, some of which consisted of shaking it and nodding it rather violently in an attempt to challenge his system, he showed no signs of vestibular problems or disorientation. The only complaint that he had—other than a minor irritation of the skin where the biomedical sensors had been pasted on—was that just before he was picked up at sea he had experienced a "mild sensation of stomach awareness." The sensation had passed in a little more than an hour, and the doctors made it clear that it "in no way approximated nausea or vomiting." The temperature outside the capsule when he landed was 76 degrees Fahrenheit. The temperature inside the cabin was 103 degrees. The temperature inside his suit was 85 degrees, and the relative humidity of the outside air was 60 to 65 percent. Glenn was uncomfortable, but he was not seasick. The doctors also checked out Glenn's blood count both before and after the flight. With the same technician making the tests, his red blood count a month before the flight was 4.75. A week before the flight it was 4.96. Eight hours after the flight the count was 4.82. And 46 hours after the flight it was 5.03. His hemoglobin count for the same periods of time stood at 14.5, 14.1, 16.1 and 14.7. Other tests broke down the exact content of glucose, potassium, albumin, calcium, chloride and various plasma enzymes in Glenn's blood for comparison with similar analyses of his blood that had been made before and after centrifuge runs and various preflight activities before the orbital mission.

One of the doctors who was waiting for me at Grand Turk was our staff psychiatrist, George Ruff. He gave me a form to fill out that I had already filled out at least a hundred times before and after all of our training missions and dry runs. The last question on the form asked: "Was there any unusual activity during this period?" The question was designed to catch us up on anything that had worried or bothered us during training so the psychiatrists would know how we had reacted. I could not resist my answer. "No," I wrote, "just the normal day in space."

We had time to read some newspapers on Grand Turk while we were taking our various tests and making out our detailed reports. I began to get an inkling of the impact which the flight had made back home and the way people seemed to feel about it. I had known all along, of course, that the pilot of a successful orbital mission was going to get quite a bit of attention. But I could not possibly have been prepared for the tremendous reception which we received when we arrived back in the U.S. It seemed as if this national accomplishment had stirred the nation's pride. As the focus of that pride, I felt overwhelmed. As we went through the parades and faced the great crowds which came out to see us in Florida, Washington, New York and my home town of New Concord, I looked at all of those thousands of faces and waved back and smiled at them when I really felt much more. Here were all of these people, identifying themselves with me; and here I was—identifying myself with them right back. It was a great experience for all of us in the program to see that pride in our country and in its accomplishments was not a thing of the past. It made us all feel good that our teamwork had paid off and that it seemed to mean so much.

I meant what I said at the time about being but one member of a great team of Astronauts and technicians who had made this flight possible and who would now move on to bigger plans and longer flights. I also meant it when I pointed out that we had only scratched the surface of space. We still have a long way to go before we unlock any of the great secrets of our universe or make any of the really important discoveries that a man can uncover in this new environment. It encloses us all and affects all of our lives, so we must push on and find out more about it.

My flight proved a number of things. Perhaps the most important lesson is that man belongs in space. Without him, the spacecraft is deaf to the sounds and blind to the sudden problems and opportunities that can present themselves. The systems will change as we develop new and better ways of doing things. The systems we have used in Project Mercury in order to gain a toehold will give way to other schemes. No matter what their design is, however, man has the reliability to operate the man-spacecraft combination and the adaptability to insert himself into the system and make it work. We never did consider the Astronaut to be merely a passive passenger in Project Mercury. Now we know that man has a key role. Man is not infallible, of course, but on future flights he will be less dependent on the machine itself. On the longer-range flights to the moon and beyond he will be even more on his own, and will be required, for example, to make repairs to keep the spacecraft systems operating and to make decisions affecting his mission which could not be made for him on the earth. It is a source of satisfaction to me that I was able to confirm this principle in a small way.

All of us have tried to make these same points in this book. We are only the pioneers of this new age. Other men, younger and perhaps wiser than we because of our experience, will pick up the baton and run further. We only hope that we have helped to show the way, and that the spirit of teamwork and cooperation which we have tried to maintain will guide the young fellows who are coming along after us at this very moment. At the risk of repeating myself, I would like to close this chapter—and I know that I speak for all seven of us—with the same words I used to thank the Congress for the honor it paid to us after my flight:

We are all proud to have been privileged to be part of this effort, to represent our country as we have. As our knowledge of the universe in which we live increases, may God grant us the wisdom and guidance to use it wisely.

THE CON

Malcolm Scott Carpenter

FIRMATION

John Glenn's flight in Friendship 7 was the pioneer mission, and it set a tremendously high watermark. It also made it much easier for the rest of us in a way, because it solved so many problems that had concerned some of our people and helped dispel whatever doubts anyone might have had that we were on the right track. The very fact that John had serious trouble on his flight but went on to overcome it nailed down two extremely important points for us. First, it proved that man does belong in space and that a human being who has stamina, intelligence and curiosity is the best of all possible instruments for bringing back knowledge of our environment. The flight also proved that the systems we were considering to help conquer space were reliable and sound—provided there was a man aboard to make them work. John settled any doubts that we might have had about going back to the drawing board before we could continue our plans for sending men deeper into space; we simply had to go on making fixes in the system, just as we had all along, and keep pushing ahead.

That was my mission, really—to give the program as hard a push as I could. I would give all the systems another whirl, to find out how they functioned in another man's hands and to determine if the fixes we had made were any good. But mainly, since I felt that John had been unduly handicapped during his flight, I was most anxious to let the capsule take care of itself as far as possible and concentrate on things that he had not been able to do. I was to have more freedom to measure, study and observe events *outside* the capsule. And I had many sciences to serve. I had a new kind of camera, some special film, and a whole glove compartment full of instruments for making measurements of the sun, the stars, the horizon and those particles that John had seen. (We fixed the problem of the overflowing ditty bag that John had had by installing a compartment on the instrument panel to hold our loose gear during flight. Since it looked a lot like one, we called it our glove compartment.) The weather people wanted me to make further observations of cloud formations and of a pilot's ability to study them from our orbital altitude. Our own space scientists wanted us to make pictures of the horizon on special film through a special filter which would help them to work out devices for future spacecraft that would utilize the horizon as a reference point for navigation during lunar flights. Because of all this extra work, I was much more interested in being a kind of prospector on my flight, I guess, than a pioneer. John was the pioneer. But he did not have the time or opportunity to bring back many riches. I wanted to mine the heavens and return from my flight with as much scientific information as I could.

Despite these objectives, I was tense and not at all at ease during the days immediately before the launch was first scheduled to go. I did not like the waiting, and I felt I needed more time to work on details of the flight plan and to work with the special equipment I would be using. I was not convinced in my own mind that I would perform well under the stress of the flight, and one night I had real difficulty getting to sleep. I suppose that fear had something to do with it. I was not afraid for myself, but I knew it was a dangerous mission and I hated the thought of losing the life of the father of my four children or of missing out on experiences that I looked forward to so much. I was rather sensitive about the company I kept at this point; there were only a few people I really wanted to be with, and when I was not working I remained either alone or in the company of John Glenn and a few other close friends.

Then the flight was postponed, and things took a turn for the better. The scrub gave me a chance to practice more with the flight equipment and to study up on a few things about the flight that I had been worried about. I built up confidence, and I began to think again the way I had been thinking for three years—that the flight would be a magnificent experience for me and that if it was successful I could make a valuable contribution to the program. I ate well and slept well. As the new launch date came closer I kept waiting for the tenseness to return, but it never did. I had finally reached the crest of the hill. Everything came out even, and I felt for the first time that I was really a part of the machine.

It is rather ironic, I think, that I could have taken off for space right on time that morning if it had not been for such a mundane thing as a brush fire in the Everglades. There was a pall of smoke and haze over Canaveral and we held for 45 minutes for it to clear before we could go. I would not have called my wife, Rene, from the capsule if it had not been for this hold. She and I had agreed that such a last-minute conversation might be upsetting for both of us. And up until then the countdown had been moving along so perfectly that I had too much to do to take time out for a call. But during the delay for weather I decided that I wanted to call and that I could handle it. Rene was having breakfast with the children in a house about 10 miles down the beach from where I lay in the capsule. In fact, when she answered the phone, she told me she had her mouth full of Wheaties. She said she could not see the Cape yet because of the haze. I talked to Jay and to Scottie, who told me he knew everything would be all right. Candy informed me that she had just gotten up. And Krissy asked me how the test was coming along. I told her we were doing fine. It got through to me only once, briefly, and tears came to my eyes. But it passed and I went quickly back to work.

I was amazed at my own calm. I felt a certain detachment, as if I could stand a little to one side and watch myself get ready. I was more curious than anything else; I could hardly wait to see how it would turn out. Perhaps this detachment is a defense against fear, much the way shock is a defense against pain. I remember from childhood that when my grandfather was dying of a stroke, he said to his doctor: "At last I'll know the great secret." As I lay there on top of the Atlas, I was confident that everything was going to be all right. But I felt that I, too, was going to be let in on a great secret, and that this fantastic experience I had looked forward to for so long would soon be here.

The launch was just a snap. It was the shortest five minutes I had ever experienced. We rose with very little vibration inside the capsule. In fact, the ride up was much gentler than I had anticipated. The engines made a big racket, but there was no violent trembling of the whole structure. I remember thinking to myself as I watched the altimeter needle wind up to 70,000 feet, then 80,000, then 90,000, "what an odd place to be, and going straight up"—and I had about 560,000 feet still to go. I did notice a distinct swaying of the machine as we climbed. John Glenn had reported this on his flight, and he had said it felt to him as if he were bouncing back and forth on the end of a springboard. It did not feel that way to me; it just seemed that we swayed off to one side and stopped abruptly, then swayed back to the other side and stopped again. The motion was not alarming. And there had been no problem when I went through the area of "maximum q" where the aerodynamic forces piling up against the capsule had reached their peak. There was a barely perceptible vibration, but no more than I had experienced on lift-off.

The first thing that impressed me when I got into orbit was the absolute silence. One reason for this, I suppose, was that the noisy booster had just separated and fallen away, leaving me suddenly on my own. But it was also a result, I think, of the sensation of floating that I experienced as soon as I became weightless. All of a sudden, I could feel no pressure of my body against the couch. And the pressure suit, which is very constricting and uncomfortable on the ground, became entirely comfortable. The pressures were all equal; even a change of position made no difference. It was part of the routine to report this moment to the ground as soon as it came. It was such an exhilarating feeling, however, that my report was a spontaneous and joyful exclamation: "I am weightless!" Now the supreme experience of my life had really begun.

I turned the capsule around so that the blunt end behind me would be pointing along the track that I would follow. I used the manual control system to execute this maneuver, and it worked perfectly. Then I checked out the entire manual control system thoroughly and found that the capsule responded like a jewel. Each time I moved the stick, the small thrusters on the outside of the capsule pushed out their jets of hydrogen peroxide steam and moved Aurora 7 into whatever position I wanted. Then I checked the automatic control system and discovered that it did not seem to be working so well. As nearly as I could tell, the system was not aligning the capsule properly at all. There seemed to be a constant error in the horizon scanners that sense the position of the horizon by detecting the narrow line between the warm earth and the cold sky. The error was in pitch—I was pointed lower than I should have been. But I had a lot of things to do and a limited amount of time in which to do them all, and I decided there would be time to figure out that problem later in the flight.

I had looked out of the window briefly when the escape tower was jettisoned, and I caught a glimpse of it right on the horizon, streaking away like a scalded cat. But in the early part of the first orbit I concentrated mainly on the control systems and did not really look around. When I finally did, the sight was overwhelming. There were cloud formations that any painter could be proud of—little rosettes or clustered circles of fair-weather cumulus down below. I could also see the sea down below and the black sky above me. I could look off for perhaps a thousand miles in any direction, and everywhere I looked the window and the periscope were constantly filled with beauty. I found it difficult to tear my eyes away and go on to something else. Everything is new and so awe-inspiring that it is difficult to concentrate for very long on any one thing. Later on, when I knew that I was returning to some wonderful sight that I had seen before, I could hardly wait to get there. Using the special camera I carried, I took pictures as fast as I could, and as I raced towards night at 17,500 miles an hour I saw the beginnings of the most fantastically beautiful view I have ever had—my first sunset in space.

Crossing Africa, I looked in the periscope and saw the shimmering sights disappear and the earth blackening behind me. Through the window, I could see the sun actually dropping towards the western horizon. Right on the horizon as the sun fell, a band of color stretched away for hundreds of miles to the north and south. It was a glittering, iridescent arc composed of strips of colors ranging from yellow-gold to reddish-brown, to green, to blue and then to a magnificent purplish blue before it finally blended with the black of the sky. The colors were all sharply separated and glowed vigorously, alive with light, and I watched the band narrow until nothing was left but a rim of marvelous blue. It occurred to me that I now knew what a new earth must look like from the moon (with the earth in line with the sun and lighted from behind)—like a bright blue ring in the sky. I looked in the periscope soon after that and saw nothing but blackness ahead of me. I was on the dark side of the earth.

It was on this first pass across the dark side of the earth that I used up a lot of my fuel. I kept trying to move the capsule around from one position to another so I would not miss anything and so I would be in a better position to take pictures. And then, when I got over Australia, I was supposed to watch for some flares that were being set off on the ground to see how clearly they could be observed from space. Unfortunately, Australia was mostly covered by clouds and I did not see the flares. But I maneuvered the capsule around a great deal trying to find them, and this was a costly bit of travel, too. It is possible to change the capsule's attitude gradually just by setting up a gentle movement with the controls and then letting it drift into the desired position. This is the most economical way to do it; you save fuel this way. But it is also a slow, time-consuming procedure and I was impatient to complete all of the items on the flight plan. So I kicked the capsule around faster by using up more fuel and pushing it all the way. It was an expenditure that I would regret later on.

I also began my observations of the night sky at this time. The stars were bright, and though I do not believe I saw as many of them through the window as I might have seen on a clear night down on earth, I found that I could track them easily. I discovered that I could hold the capsule in the proper attitude just by fixing on a known star near the horizon and keeping it centered in the window. Then, as the sun rose ahead of me, I got my first look at John Glenn's "fireflies." As they drifted around the capsule near the window they looked more like snowflakes to me, whitish in color and varying in size from one-sixteenth to one-half an inch in diameter.

I was already having some trouble with the suit temperature at this time. I was hot and uncomfortable, and the sweat was pouring off my face and running into my eyes, making it difficult for me to see things properly. I tried various settings on the valve which controls suit temperature; and the ground control stations made other suggestions. But nothing seemed to work particularly well. The heat itself was not bad—the temperature in the suit never rose above 84 degrees—but the humidity was not being controlled as it should have been. In fact, I found that when I put up the visor on my helmet and exposed my face to the much hotter atmosphere of the cabin—which was over 100 degrees—I got the same kind of relief one would get if he stepped out of a stuffy room into the fresh air. This was because the cabin air was far drier than the air in my suit. All in all, it was a bothersome problem, and fussing with the controls took up more of my increasingly tightening schedule. (The sweating also explains most of the seven pounds that I lost between launch time, when I weighed 154, and postrecovery, when I weighed 147.)

I crossed the United States in early morning light. The ground seemed closer than I thought it would, and though it was 100 miles down and I was going 5 miles per second, it seemed to pass underneath me at about the same speed as in a jet at 40,000 feet—perhaps just a little faster. Where it was not covered by clouds, I could see the ground remarkably well. I could note rivers and lakes—and even a train on a track. And as I passed over farm country in the southwest I could tell where the south 40 was cultivated and the north 40 was lying fallow. At every new sight, my elation was renewed, and I kept waiting again for the next one.

As I passed over Cape Canaveral the first time around, I started the balloon experiment. It did not go according to plan. I deployed the balloon, which was made of a combination of synthetic fabric and aluminum. This was an important experiment, to measure the drag of the balloon in the very thin atmosphere and observe its behavior, its distance from the capsule and the various colors it was painted. But the balloon did not inflate properly—it got only about 10 inches wide instead of 30—and it took longer than I had expected for it to get to the end of its 100-foot nylon tether. I had to do some fancy maneuvering with the controls to keep it behind me and to prevent it from wrapping itself around the capsule. It did not oscillate as we had expected it to. Its motion was completely random in yaw and pitch, and when it reached the end of its line it would bounce back and the line would get slack and kink up again. I was able to judge its colors—I think the bright orange was most visible, which is a clue we may use for painting objects that we want to find and link up with in space when we start our rendezvous procedures. But it was nearly impossible for me to make any drag measurements. And finally, it refused to jettison when the experiment was over. The switch that was supposed to release the balloon did not operate, and I continued to trail the balloon until retro-fire, like a tin can attached to the rear bumper of a car. All in all, the balloon was sort of a mess.

I was appalled at the end of the first orbit at the low state of my fuel. I had not much more than half my supply left for both the automatic and the manual control systems. And I was warned by the Control Center that I would have to come down at the end of the second orbit if I was not more careful about conserving fuel. Cutting the flight short seemed like a terrible prospect to me. There was so much to do and see that I needed all the orbits I could get. I made up my mind to be very careful with the fuel from then on, and for the rest of the mission—until it was time for retro-fire—I did quite a bit of drifting. I would start the capsule in one direction with a slight movement and just let it sweep around on its own. I was able to make quite a few observations this way—though not so many as I could have if I had had more control—and I was impressed again and again with the wonder of weightlessness. A change of attitude means nothing in this state. Everything floats. At one point, a washer that had come loose somewhere went floating past my face. I picked it out of the air and put it in the glove compartment. It leaked out two or three times after that. Once it just hung in front of me for two or three minutes and didn't move an inch. And later on in the flight, when I had nothing more pressing to do at the moment, I started experimenting with the camera. I tried to see how still I could get it to stay out there in front of me. I'd give it a little shove with one hand and stop it with the other. I bounced it off my fingertips and tried spinning it in front of my face.

The eating problem was not quite so amusing. I had been provided with some special bite-sized squares of food—cookies, chocolate, pudding and date bars—all covered with a white waxy substance so they wouldn't crumble. But they did crumble, all over the place. Every time I opened up the bag, the crumbs would come crowding out like a swarm of bees, and I tried to pick them out of the air and eat them to keep them out of the cabin atmosphere. One recommendation I made when I got back was that we be provided with a transparent bag for food on future flights so the pilot can see what he is reaching for and pick the pieces that haven't already started to fall apart.

Midway through the second orbit, the suit temperature got quite bad. The ground was reporting that my body temperature was being recorded as high as 102 degrees. I did not think this was correct. I did not feel that I was actually running a fever; but I did begin to notice one of the first symptoms of a fever. I was having trouble finding the right words with which to express myself in my reports to the ground. I knew what I wanted to say, but I could not seem to say it with the same fluency I had felt earlier in the flight. This worried me a little. My main concern was that I was going to be hot enough during re-entry when I ran into the heat pulse, and that I wanted, if anything, to be too cool by that time rather than too hot. I wanted to go into re-entry with a cold soak.

I tried very hard to think the problem through. "Let's just back off," I thought, "and analyze this thing carefully. Are you too hot or not? Are you thinking clearly? Is there really something wrong with the suit cooling system? Which way *should* you turn the valve, up or down?" My answer to the last question was that I wasn't sure at all. The people on the ground tried to help, but this was a little bit like having someone sit across a room blindfolded and tell someone else how he's doing as he combs his hair. The man with the dials has the best feel for it, and I just had to work this out for myself over a period of trial and error. After studying the problem for a good while, I came to a couple of important conclusions: First, it took longer than I expected for a change in the valve setting to affect the temperature of the suit. (This resulted in another recommendation that I brought back with me—that we ought to devise a better water-absorption device for the suit system to keep the humidity under control.) And second, I discovered that if the faceplate on the helmet is left open, the heat of the cabin gets into the suit and acts against any cooling effects of the system. That seemed ridiculously obvious, but the simple fact was that in all our preflight studies we had not taken this possibility into account. So I closed the faceplate and managed to find a setting, finally, that made me comfortable. I was in good shape for the third orbit.

On the last time around I did a lot more drifting. Although I made several attempts to figure out what was wrong with the automatic control system. I was not able to come up with a solution. I found out later that the two horizon scanners which gave information on the capsule's attitude to the automatic control system were apparently not working accurately. I did not know this at the time, however, and I used up a good bit of fuel trying to straighten things out. Then I got involved in other matters. The earth and sky were still waiting for me to investigate them. It seemed a pity that I was having to spend so much time worrying about a man-made object when God's own creations, just outside the window, were much more mysterious and challenging.

The last hour before retro-fire passed quickly, just as all the rest of the time had. Flying through space, I felt a curious compression of time, as if the speed at which I traveled had some effect on the length of moments I spent there and packed them too tightly on top of each other. I always seemed to be in a tremendous hurry to get from one event to another, as each new sequence popped up like a duck in a shooting gallery. I photographed my last sunset, tracked my last stars, ate a chocolate bite and some cookie crumbs out of the air and drank quantities of water through a rubber tube in an effort to keep myself cool inside the suit. I was doing something all the time.

I was really busy when I got around to Hawaii for the last time. I was maneuvering the capsule around to make pictures of the sunrise; I was trying to stow equipment away so it wouldn't bounce around during the re-entry; and I was talking to the Hawaii tracking station about my retro-fire procedures. And then, suddenly, one of John's fireflies came by the window again. It was a particularly bright one, and I reached out to grab a light meter to take a reading on its intensity. As I did this, I hit my hand against the wall of the cabin and a whole cloud of particles flew off past the window. I was fascinated by this surprise, and I started thumping the wall all around me. Every time I hit it, more particles popped away. Surely, I thought to myself, they must have been clinging like frost to the capsule and were coming from the capsule after all and not, as John had thought, from some other source. There wasn't time to think about it any longer, for it was time to get ready for retro-fire and the long return to earth. But I believed that I had solved the mystery: they were bits of frost that had collected on the cold outer surface of the capsule.

The flight plan called for me to use the automatic control system to get the capsule into the proper attitude for firing the retro-rockets. They had to go off right on time, and the capsule had to be aligned correctly before they fired or I would not start down at the right angle and land in the right spot in the Atlantic, more than 3,000 miles away. I tried using the automatic controls. The big thrusters spilled out the precious fuel, but they still did not bring the capsule around to the correct attitude. So I switched to the fly-by-wire mode and tried this combination of automatic fuel and manual controls. My manual tanks were now very low. And here I made a mistake. When I switched over to fly-by-wire, I neglected to shut off the straight manual system. As a result, I was draining fuel out of both systems every time I used the stick.

At the time of retro-fire, I believed that I had brought the capsule to the proper attitude. I found out later that this was not correct. The small, bottle-top end of the capsule which was trailing me was canted 25 degrees to the right of where it should have been. In other words, I had an error in yaw. I had just not been able to line the capsule up on all three axes as precisely as I should have. This meant that the capsule was not pointed in an absolutely straight line along its path when the rockets fired, and so it did not slow down as much as it should have. This was mostly my fault, though it is difficult to judge yaw visually by day when you have no reference point outside to guide you, and it accounted for about 175 miles of the 250-mile overshoot. But a couple of other things went wrong in addition to that. First, the retro-rockets did not deliver the full thrust that we expected of them. John reported that he felt as if he were going back to Hawaii when his retros fired. I did not feel the same sensation at all. The firing was comparatively gentle. This loss of thrust accounted for another 60 miles of the overshoot. And then, on top of all this, the three retros fired about three seconds late. They were supposed to fire automatically, but they did not. I watched the clock pass the correct instant, and then I hit the retro-button myself a second later. Two more seconds passed before they finally went off. I noticed a puff of smoke in the cabin at this point. It smelled like hot copper, and was probably caused by a short circuit in the retro-firing mechanism. I was not so concerned about a fire, however, as I was about the delay in the firing. At my speed of 5 miles per second, this lapse of 3 seconds accounted for another 15 miles in the overshoot.

In between the time when I fired the rockets and the moment that Aurora 7 began its re-entry through the atmosphere, things were pretty tight. My fuel supply was critically low, and I was not at all sure that I had enough left to keep the capsule in the proper trim for the long glide back to earth. If we came through at the wrong angle and the fuel was exhausted, I would be unable to control the capsule during descent and the chances of surviving such a re-entry were not good. If the capsule was lopsided during re-entry, one side of it could overheat, and the heat could even eat through the thin upper wall and destroy the parachutes that I would need to get down. I have listened to recordings of my voice which were made during this tense moment. I don't sound confused or frightened in them, but strangely dejected. The tone of my voice dropped measurably, and all of my sentences ended on a minor key. But, even at this low point, I remember thinking, "This has been the greatest day of your life. You have nobody to blame for being in this spot but yourself. If you do everything correctly from now on, you may make it. If you do not, you just won't." I decided I had to use more fuel to stay in the proper attitude. But now I found that though the manual tank still registered 7 percent, it was really empty, and I had only 15 percent of my supply left in the automatic tank for the whole re-entry. I was dangerously short.

I maneuvered the capsule very gingerly, keeping the horizon in view through the window, and trying to use as little fuel as possible. I held my position steady, and when I felt the first welcome oscillations that told me the capsule was encountering the heavy atmosphere, I started the capsule rolling at a rate of 10 degrees per second. This was to help keep it on its proper course on the way down—and equalize the heating—and we were headed down now.

It was actually a beautiful re-entry despite all my worries. The ride

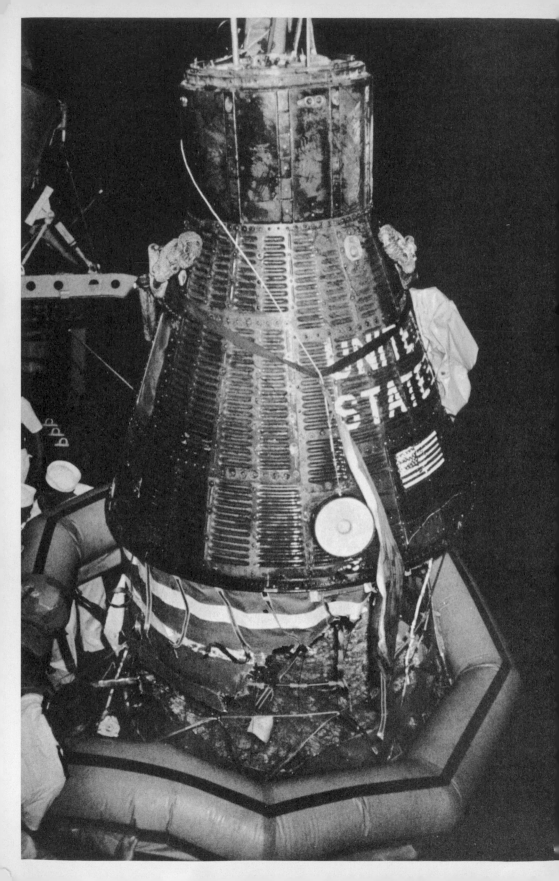

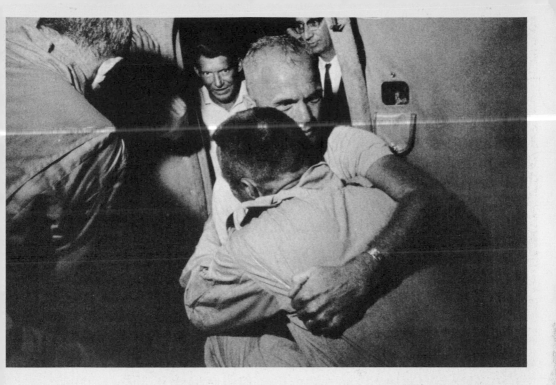

Scott Carpenter was the fourth American to go into space, but because of the long and unexpected uncertainty about his safety, emotions were as strong when he got back as they were for any of his predecessors. The destroyer crew was jubilant as it lifted his capsule out of the water and onto the deck [OPPOSITE]. The medical officers on board the recovery ship sent admiring glances as Carpenter walked a narrow rail in a post-flight test of his balance [RIGHT]. John Glenn had a warm embrace for his colleague when they met for the first time after the flight [ABOVE]. And Scott was a proud and thankful father as his children greeted him on his return home.

most of the way down was perfectly smooth, and we headed in at a good angle. When I glanced out the window I could see an orange doughnut of fiery particles stretching out like a wake behind me. These were tiny bits of the ablative heatshield which had melted off and were carrying some of the intense heat away with them. Everything was normal. Then I noticed a green, hazy glow building up around the narrow end of the capsule. Since this is where the parachutes are stored, I was worried again that perhaps they were being damaged by the heat. I spoke into the tape recorder as I watched the glow, so there would be a record of all this if anything did happen to me. "I see a green glow from the small end," I said. "It looks almost as if it is burning away. Oooooooh, I hope not." As I spoke these last words, I had to laugh out loud. That was one of the lines that José Jiménez uses in the Astronaut record that we had all had so much fun with, and it amused me that the joke had come out of me, just like that.

The peak Gs lasted longer than I expected them to on the way down, and as I talked into the tape recorder I had to inhale more frequently in order to speak. This was the only time on the mission that I was really aware of the G forces. I was explaining everything I saw—the flaming particles, the orange glow and the green haze from the hot beryllium shingles on the top cylindrical section of the capsule. I was never frightened, just interested, during this phase of the flight. The oscillations were building up now and I could feel them and hear them going "bang, whump, bang, whump," as the capsule swung from side to side. They were welcome, because they meant that an aerodynamic force would be exerted against the capsule and help keep it on an even keel on the way down. The G forces tapered off now, at about 120,000 feet, and the capsule and I were falling about 600 miles an hour. But the swaying built up rapidly, and I used the very last of my fuel trying to control it. I was concerned that the capsule might topple over completely and start coming down topside first. If this happened, the drogue chute could get badly fouled up if it came popping out during this wild swinging, or it might snap the capsule around so violently that the chute would be badly damaged during deployment.

I was still not frightened. It was a tight situation, and I was very alert. There just isn't time in a situation like this to get wide-eyed and ask yourself, "What'll I do now?" You don't clutch because you can't—that *could* be sudden death. You just have to keep interested in what is going on and work your way out of it. Finally, as the oscillations got worse and the capsule started to sway through a huge arc of about 270 degrees—almost full circle—I punched the button to deploy the drogue chute. This was at 26,000 feet. The flight plan called for me to punch it at 21,000, but I felt I needed it sooner to help damp the oscillations. The six-foot drogue came out in good shape, and the descent stabilized. I had been in a layer of haze, but I was falling through clouds now. The altimeter swung towards 10,000 feet, the point at which the main chute was supposed to come out automatically. When it did not, I gave it 500 feet more and pulled the ring. Out it came, a glorious orange and white canopy, perfectly shaped, stressed to its limit and drawn taut as sheet metal as it strained to support the capsule's weight.

I did not know it just yet, but the long period of suspense was beginning now. I was not concerned because I had no way of knowing inside the capsule that I was overshooting the target area. We had experienced the normal communications black-out during re-entry as the ionization barrier built up around the capsule, and neither the Cape nor I could hear each other. On the way down in the chute, I picked up a transmission from Gus Grissom back at the Cape. He was transmitting blind, but he advised me that I was long and that I should expect to wait for about an hour on the water before recovery. He also said that a plane carrying paramedics was on its way to the landing area to give me assistance. The tracking devices had computed my landing point as I came down, so the Control Center knew fairly well where I was, but it was clear that I had overshot by so far that I was out of range of our communications network. Most of our communications between the capsule and the ground were made on a line-of-sight basis—that is, so long as I was at orbital altitude, the radio transmissions carried easily to the next tracking station. But the lower I got, the shorter my range of communications became until when I reached parachute level at 26,000 feet, there was no one close enough to hear me. I did pick up signals from the stronger ground transmitters, but mine were too weak. I made several calls myself as I parachuted down, but when I got no answer I knew that no one could read me. I was beneath the clouds now and I could see the water. I got set for the landing.

We did not hit hard at all. It was almost like sitting down in a chair. But the capsule went completely under water and came up listing quite sharply—about 60 degrees—to one side. I could see a little water in the cockpit; the tape recorder down by my feet had several splashes on it. All things considered—especially the hour or so it would take them to reach me—I thought it would be sensible to get out and wait in the raft. I took off my helmet, removed the right half of the instrument panel to make an exit and then squeezed my way up past the instrument panel. It was not easy, but it was better than sitting inside a listing capsule or blowing the side hatch and losing the capsule. I opened the hatch on the small end of the capsule, put the camera I had been using in a safe place near the opening and dropped the life raft into the water. I got into it before I realized that it was upside down. I climbed out into the water, turned the raft over and got back in. Then I tied the raft to the capsule so we wouldn't drift apart, and turned on the SARAH beacon which would help the recovery aircraft home in on my position. And then I said a prayer: "Thank you, Lord," and relaxed for the wait. I have never felt better or happier in my life. I felt like a million dollars. I had gotten a little water into my suit when I jumped into the ocean to right the raft, but this was not unwelcome because it kept me from getting too hot inside the suit.

I sat for a long time just thinking about what I'd been through. I couldn't believe it had all happened. It had been a tremendous experience, and though I could not ever really *share* it with anyone, I looked forward to telling others as much about it as I could. I had made mistakes and some things had gone wrong. But I hoped that other men could learn from my experiences. I felt that the flight was a success, and I was proud of that. I had brought back a number of new facts. I had made what I felt were some good intensity measurements on the stars—particularly when they pass through the luminous layer just above the horizon that John had reported on during his flight. I had made some measurements on the width of the layer—its distance from the earth—and how much stars are dimmed as they pass through it. By pinning down the time it takes a known star to pass through the luminous layer and reach the actual horizon, we could work out a way to compute our position in space from the occlusion of a star. I had used an air-glow filter to determine the exact frequency of the light waves that emanate from the layer itself. I had stumbled across what seemed to be a logical answer to the mystery of the fireflies, so we could probably cross that one off our list and get on to other mysteries. I had satisfied myself that we could orient in yaw with the stars at night, and that drifting flight without controls would offer no problem. I had confirmed John's findings that weightlessness was no problem—at least not on a flight as short as ours were. My co-ordination had not been affected in the least by the weightless state; I had given myself several tests to prove this—including closing my eyes and reaching out with my arms for specific spots on the panel—and I had had no trouble going right where I wanted to go. If anything, the mission had seemed much less rugged than John's. Everything about it—the Gs, the retro-firing, the re-entry—had apparently been gentler. It was not in the least stressful. There was no real effort to it. You just worked mentally, that was all.

I also brought back some ideas for improving the systems before the next man went up. Aside from recommending a better temperature control for the suit and some way to block off the large thrusters on the automatic control system so we don't spill so much fuel out making fairly minor corrections, I had a number of other suggestions for the engineers and I went over some of these in my mind as I sat in the raft. One mental note I made was that there should be a signal on the instrument panel to let the pilot know when his faceplate is open and when it is closed, so he does not forget in the midst of all his other activity and leave it open when it ought to be closed for safety's sake.

For a long time as I thought about all this, I looked at nothing but sky and sea. I saw a patch of Saragasso weed go drifting past. Then I noticed a black fish about 18 inches long hanging around near the raft. He was friendly and tame as a chicken and so close I could have reached out and grabbed him. But I didn't because that might have hurt him, and at the time he was my only friend.

A little later I heard airplane engines, and saw the first P2V patrol plane approaching. I signaled him with a hand mirror, and he began to circle my position. I knew he had seen me. Not long after that it seemed to me that there were planes all around me. There were so many of them, in fact, that I did not notice it when the first paramedic jumped into the ocean. The first indication I had that he was even around was when he swam up behind me, grabbed the raft and said hello. We started up a normal conversation just as if we had happened to meet on some street corner. The second man came up right afterwards, and they both got busy attaching a collar to the capsule to keep it afloat. These men probably saved the capsule from sinking, and they had no way of knowing that I did not need medical help. So I should have been very grateful for their presence. And I did thank them and offer them some of my food and water. But in a funny way I resented the fact that they were here. This was my part of the ocean, and I wanted to be there alone for a while longer and contemplate what had happened. It was still going through my mind. But now I had a new train of thought. I had not really savored the mission until now. In fact, all the training and preparations for the flight had been so intensive and tiring that not long before I went I felt that I would never want to go through this again.

I certainly did not feel like that now in the raft, however. I felt that space was so fascinating and that a flight through it was so thrilling and so overwhelming that I only wished I could get up the next morning and go through the whole thing all over again. I wanted to be weightless again, and see the sunsets and sunrises, and watch the stars drop through the luminous layer, and learn to master that machine a little better so I could stay up longer. There's no doubt about it, space is a fabulous frontier, and we are going to solve some of its secrets and bring back many of its riches in our lifetime. I would not miss that for anything.

INDEX